Evangelical Anxiety

Evangelical Anxiety

A Memoir

Charles Marsh

HarperOne
An Imprint of HarperCollins*Publishers*

EVANGELICAL ANXIETY. Copyright © 2022 by Charles Marsh. All rights reserved. Printed in the United States of America. No part of this book may be used or reproduced in any manner whatsoever without written permission except in the case of brief quotations embodied in critical articles and reviews. For information, address HarperCollins Publishers, 195 Broadway, New York, NY 10007.

HarperCollins books may be purchased for educational, business, or sales promotional use. For information, please email the Special Markets Department at SPsales@harpercollins.com.

FIRST EDITION

Designed by Kyle O'Brien

Library of Congress Cataloging-in-Publication Data is available upon request.

ISBN 978-0-06-286273-0

22 23 24 25 26 LSC 10 9 8 7 6 5 4 3 2 1

For K

Maggie, you will be thrown into the world someday, and then every rational satisfaction of your nature that you deny now will assault you like a savage appetite.

George Eliot, *The Mill on the Floss*

Something has happened to me, I can't doubt it any more. It came as an illness does, not like an ordinary certainty, not like anything evident. . . . And now, it's blossoming.

Jean-Paul Sartre, *Nausea*

Contents

CONTENTS

PART V | After Analysis

PART VI | Quiet Days in Charlottesville

PART I

Life's Rich Pageant

Martin Luther on Prozac

Sometime after my family moved to Charlottesville, on a mild April evening, I sat in my accustomed pew beneath the Tiffany stained-glass window at a downtown church. The church had recently launched an evening service intended for students and seekers, with tasteful acoustic praise songs and a fellowship supper in the refectory afterward prepared by a local farm-to-table caterer.

That evening, I was treated to a sermon on the sixteenth-century German Reformer Martin Luther that included this pastoral word to the people: "It is my hope and prayer that every person in this congregation suffer a complete nervous breakdown before they reach middle age." He may have said before their thirtieth birthday. I can't recall. I had recently rounded forty but didn't think of myself as middle-aged. I'd only been teaching for a decade, free of doctoral studies and gainfully employed, and I could still run with the kids at North Ground gym in afternoon pick-up games.

The preacher, an amiable fellow a few years younger than me, had briskly approached the pulpit in crisp khaki trousers and a Vineyard Vines button-down and begun the homily by surmising that if Martin Luther had taken antidepressants to quell his violent mood swings and inner torment—"I am like ripe shit, and the world is a gigantic asshole," Luther might bark out—the world would have never known a Protestant Reformation. Its great insights—justification by faith alone, *sola scriptura*, and the priesthood of ordinary believers—would have been lost to recalibrated serotonin levels and steadier nerves.

It was a sermon I knew I would not soon forget; the gospel of the shattered self, where the anxiously ill submit graciously to the whip. By the time the priest concluded that "our lives are not a journey, they are a train wreck," air-quoting "journey" to scattered laughter, I felt the old rumblings of dread begin their slow rise to a symptom.

The whisper-folk benediction be damned, I whisked my wife and children out the side exit and ricky-bobbed the mini-van the two miles back to our house and there hurtled upstairs to my medicine bag, shook free a lorazepam, and placed the round white pill under the tongue. "The body of Christ broken for you," I thought, waiting before the bathroom mirror for a direct hit to the blood and the "Amen" I needed most.

Harvard Divinity School: Fall 1981

I want to talk now of an autumn evening in Cambridge, Massachusetts, near the end of the time when I lived so successfully disguised to myself as unaghast and free. It was an ordinary night during my first semester at Harvard Divinity School. I lay in bed in the low-ceilinged, overheated dorm room to which I had been assigned, when a panic overcame me so complete that from that moment on I could divide my life between all that had come before and everything that followed: twenty-three years, six months, and three days of coming to feel at home in my mind and body, changed in an instant.

I had set my oscillating fan to high and slipped into a shallow sleep. It was not unusual for me after an arousing day of study to spend an hour in bed, letting my mind roam free, resting in the silences.

The first sign that something was wrong came from a crackling sound in the drywall that separated my bed from the one in the next room, occupied by a visiting scholar from a midwestern college. I knew my neighbor already as one who also suffered from allergies and asthma. He'd been having a rough go of it lately; the building's dust mites and allergens patrolled the hallways like prison guards. It's something we'd talked about briefly, our shared miseries. My own inhaler remained at arm's reach.

I clicked off the fan and sat up in bed. It was then that a high pandemonium ripped away everything protecting me from the world outside. I was no longer a person alone in his room. In an instant, I could hear all things inside my body in their deepest repercussions. My heart and its soft aortic murmur, my breath's every exhalation and inhalation, the downward silences, the laborious intake—would this one be the last? How much noise the body makes when amped up on fear! I could hear the hiss of molecules colliding. And outside in the yellow night, the compressors harrumphing atop the nearby physics building, the sound of car engines and slamming doors. All these things I heard as tormenting assault, a soundscape I could not mute. I'd become a thought thinking about thinking itself and nothing else, metaphysics' ancient curse. A cogitation cycling through every autonomous body function, placing on each a question mark like flowers for the dead.

I wrapped my arms around my chest. Something terrible was happening. A hyperawareness of all my body was doing to keep itself alive, each system clicking to attention like some deranged mental jukebox in which I had been taken hostage. I came to understand that the sound I heard was not the strange resonance of buildings at night or the result of my nerves addled by caffeine. I heard the sound of collapse. Imagine, if you can, the skid and shatter of the barrier that protects your body from your own imagination. Imagine knowing that your body is now a park that grows predatory at sundown, like the park where the dismembered bodies of three boys were found scattered around a fire pit; I'd seen that on the television news one night when my parents were away.

I became overtaken by the harrowing notion that my body was short-circuiting, that I would soon stop breathing, and from now on every systole and diastole required voluntary operation. And I was certain that something further, deeper had changed, that some mystical unmaking had begun. What did I want in that moment? Only to quell the specter of needing consciously to operate my lungs and my heart. What did I feel? Only a totality of terror that I had never known before. I couldn't imagine a way out of the night.

———

On the first morning after the breakdown, I arose to a gray dawn metaling against the modular window. My late arrival to the seminar on "The Problem of God" was my symptom's debut in real time—it was my habit always to be present and alert when the bell rang. An angry Mennonite atheist taught the class. He spoke the way he wrote, in a wooden prose that bushhogged Christian orthodoxy to dust and ashes. And he expected his students to do the same. So in the weekly reflection papers you would see a lot of "god" and "godself," because writing "God" and its attendant pronouns "he" or "himself" puts you on the side of the divine patriarch who castrates women.* The only real referent for the word "God" is the idea we hold in our minds. The professor talked a lot about constructing the concept of God, but as far as I could see, he did not believe in the reality of God: God as being, transcendent, wholly other; God as triune; God as anything to which the Bible or creeds testified. He believed only in imaginative constructs.

That morning, in the seminar room we had been divided into groups of three and four, each assigned a doctrine—incarnation, Trinity, atonement—and tasked with recasting the Christian faith according to our tastes and preferences. Too late to join one, I labored in a solo effort over the practice of prayer. The professor, I intuited, would approve a form of speech that enabled you, when praying, to make present what is promised, the new creation, or something like that. I'd taken the idea from a Christian anarchist and tried on this day, as on all others, to strike a generous tone. Maybe I came across as self-righteous—I can assure you I felt the opposite. I'd already pitched myself as a plucky believer willing to defend Christian truth claims, and I guess I hadn't been careful to follow the professor's script.

---

* This gender-bending metaphor comes from Mary Daly, who, in *Beyond God the Father* (published in 1973, and heavy in the HDS air and on HDS syllabi during my years), wrote: "The divine patriarch castrates women as long as he is allowed to live on in the human imagination. The process of cutting away the Supreme Phallus can hardly be merely a rational affair." When Professor Daly, who taught at Boston College, announced that no male would be permitted to attend her class on feminist ethics, she raised the ire of the Boston College administration, but I was more than happy to oblige.

When I read my ode to prayer aloud (had I tipped my hand and indicated that prayer has an addressee, specifically the One who made, redeemed, and sustains the cosmos?), a seminarian from St. Louis—whom I'd known as a pleasant enough chap from meals in the refectory—slammed a fist on the table and bellowed, "That is total bullshit!" The professor's face narrowed into a thin smile while the rest of class coalesced in a giggle.

———

After class, I tried to recover my equilibrium passing an hour with my favorite companions—books, in the form of a visit to the Grolier Poetry Bookshop on Plympton Street. As I thumbed through a volume of Marge Piercy (the book felt reassuring in the hand), a woman in a plaid miniskirt approached me. She was holding a volume of Conrad Aiken, the largely forgotten poet from Savannah. I recognized her as one of the Widener Library regulars—we'd exchanged greetings a few times but never met. Now she stood beside me in this lovely little room, bending to whisper, "I miss seeing you in the stacks."

For a glorious moment she appeared to me in sunlight, enshrouded in flowers, as in a garden drawn by Turgenev. I dawdled beside her, beaming, expansive, and nervously thumbing my Piercy. I felt radiant. She felt radiant. Together, the two of us were radiant.

When she stepped toward me for a closer look at the book, I caught sight of her bare thigh just beyond her knee socks and realized I had once again overlooked, or rather tried to sidestep, the erotic element that particled through every work of art, and through my every desire.

I could have turned to her, smiled sweetly, and said, "I miss seeing you too"—or, perhaps with a little more time, "I like Aiken too, his romantic alienation, etc., and someday I'll spin a web between two dusty pine trees and hang there like a spider"—but instead I replied, "Well, thank you very much." Then I laughed out loud and made a crazy face, and dream girl exited the store and vanished forever.

Two days later, I tried again for the consolation of books. It was overcast, and I found myself at a subterranean café, sitting at a glass table with my coffee and paperback copy of Sartre's *Nausea*. Clean checkerboard tiles patterned the floor of the poorly heated room.

Sure, I should not have ordered the black coffee. I should not have smoked the dusty Gauloise. I should not have let my eyes linger on the waitress, with her beautiful bottom wrapped tightly in leather. Or her beautiful sinewy fingers. Or her taut body moving in its free operations. I should not have.

Except this, oh Lord. *Nausea*, coffee, and cigarettes on an ordinary afternoon—is that really too much to ask? People did it all the time and lived to tell. A week earlier I would have pulled it off too, smoked, buzzed, and blissed, indifferent to that creepy satyr boy scowling at me in the shadows.

Not now. Now my neural pathways were singed like Spit Devils on blacktop. I felt sad for the hands that thrashed on the table. I coughed nervously. I couldn't feel my feet. I dialed up a verse from the Psalms, but got a woman in the streetlamp's glow, her body halved by my sunken perch, waiting on the sidewalk. Her name was Cassie or Kate or Beate, and she was not a pure thought. She did not shimmer in sunlight. In her fishnet stockings and bow tie, she was the midnight, lonesome answer to my secret prayers. We would not grow old together.

Suffice it to say, I did not make much headway with *Nausea*. I crammed the novel into my backpack and hurried through the late afternoon streets toward the river.

Before the attack, I could lose myself in a book or warble merrily upon a theme. My attention might drift and stray. I might realize halfway into a book that I'd read it before, maybe more than once—and that the marginal exclamation points, question marks, *yes*es, and *ugh*s were in my hand. I might ascend from a fifty-page submersion with little recall of what I had read or who had done what. But that never mattered. Reading gave me the thing I needed most, a sense of floating and expanse.

I've often recalled the lesson my father taught me when I was five years old and still lived in Mobile, Alabama: how to break in a new book. Folding his hands around the binding as if in a gesture of sacerdotal devotion, he held the book open until the pages fell evenly divided at the middle. He pressed the two sides with his thumbs, stretching the back tenderly to its greatest strength. He continued from the middle to both sides of the book by alternating the movement from one side to the other in languid strokes until the back was relaxed from beginning to end.

In his church office, we laid out books he'd bought on trips to New Orleans, where he was still finishing his seminary degree, and brushed sealants into fraying covers and backs with the care of an archivist. In those early encounters with books, I'd felt a shift in perspective—like when I rode with my parents south on a two-lane highway that cut through miles of parched scrub pine and all of a sudden there they were, in the distance, the Gulf waters, and I felt transported.

Leaving the café for the river, did I think of my father? Did I wonder if I'd never read again? I often remember the child, spread out in the backseat of the car, one hand raised to the window, tracing the arc of the sky.

———

I'd spent the weekend before the breakdown with my fiancée on the North Shore, innocent of the foreknowledge that a mental crisis loomed. K had landed a house-sitting gig in Hamilton, sharing a room with her best friend in the sprawling single-floor home of the popular evangelical writer Elisabeth Elliot. Elliot had come to fame in our subculture for her memoir of the missionary labors of her first husband, Jim Elliot, among the Auca Indians in Ecuador. Jim's work was cut short when he suffered fatal spearing wounds and (some feared) was cannibalized, but his young widow had returned to the jungles and reaped a spiritual harvest.

Widow Elliot lived now with her third husband, Lars (the second, a Presbyterian professor of theology, had died of cancer) in leafy Updike country, but was often away speaking. Lars (about whom little is known, but whom Elizabeth once described as "a muscular Norwegian [who] is in good health") managed her career and book sales, fetched her dry cleaning, and the like, while she promoted her views on the Christian woman's obligation to submit to male authority. Elliot's books occupied a shelf of their own in my mother's modest but dearworth library.

With the trees in the wooded yard stripped bare, bracing for winter, and beside a fire, K lounged on the couch with an Iris Murdoch novel, and I cozied into Mrs. Elliot's leather lounge chair (or, more likely, her husband's, who liked his creature comforts) with Nelson Goodman's *Ways of Worldmaking*, followed by Günther Bornkamm's study of St. Paul, breaking only to jog, shower, and cook pasta—followed by a third book, just for fun, Graham Greene's memoir of his childhood, student years at Oxford, and conversion to Catholicism. I'd read *The Heart of the Matter* and *The Power and the Glory* on a family vacation to Clearwater, Florida, two years before, soaking up the gritty realism, the tortured souls counting the hours in desolate outposts, and all the other muscular obsessions that made Greene the go-to guy for disenchanted evangelicals. The memoir, honestly, felt detached and cold, and the bit about playing Russian roulette was most surely a fey Brit's bluster.

"The newspaper says the rain will continue through Wednesday," I wrote in my journal. "What is the purpose of doing theology if you deny the activity the word represents? I read Graham Greene's *A Sort of Life* over the weekend. Death, writing, failure, boredom, Catholicism, and psychoanalysis."

In college, I'd realized I was behind—not behind anyone I actually knew, but behind some notion I was forging of myself. I worked hard—and loved every minute of it—to catch up: Heidegger, Sartre, Kant's first two critiques (the second read almost devotionally), the novels of eighteenth-century England. Most of Faulkner. *The Canterbury Tales, The Rape of the Lock, The Faerie Queene.* I read them ferociously. My triumph of three books devoured back-to-back-to-back during that blissful weekend with K would mark the end of the era of immersive reading.

———

Within a week of that North Shore idyll, an ambient sadness covered over everything. Something new was taking hold.

"I was moving in a narrow range between busy distractedness and a pervasive sadness whose granules seemed to enter each cell, weighing it down," Diane Ackerman writes in *One Hundred Names for Love.* Drifting between "islands of anxiety," in a fatigue that clung to hope, "sorrow felt like a marble coat I couldn't shed."

My evangelical upbringing, as a preacher's kid in the Deep South, whispered to me an interpretation of that marble coat. I counted it all joy if I should suffer. My sorrow, my soul's sin-sickness, was not unintelligible—it was a kind of blessing, something that might draw me, like a medieval saint, to the suffering of my Lord, something that would testify wordlessly to my heroic exertion to attain purity. And, at least during those late days of autumn 1981, the heavens above and the earth below, spirit and flesh, felt miraculously aligned. Though suffering, this was the life I had craved. Or, better, in suffering—because of suffering—this was the life I'd craved.

The suffering was unfettered, alive, and I a perfect symmetry of light and heat and torque.

This attempt to receive suffering as gift was the only story available to me. I could accept my condition as a spiritual test and make the best of it, or I could . . . what? Go insane? Kill myself?

I knew the university had a student mental-health division open around the clock. A woman on my hall had gone for help after learning that her brother had died in England. I'd once overheard a doctor, a psychiatrist, I guess, speaking calmly to a disoriented student, while I waited in the adjoining room one afternoon for an allergy injection. It wasn't so much that I thought of psychotherapy with suspicion as that I'd long since formed the notion that Christians didn't need secular psychology, because we had been given the Holy Spirit, who was the ultimate Healer.

So I dug into the Word. I prayed for the mercy to endure. For the strength of days, for whatever I'd possessed the day before yesterday, for ordinary life. I prayed for a grateful heart, that I would come to accept my affliction as a gift, which meant making my life an empty vessel. I said "thank you" for the dread that covered my every waking thought, while hoping that in time this smothering weight would somehow increase my joy. In the words of the old Baptist standard: "The Lord is my light and my salvation. Whom, then, shall I fear? Whom, then, shall I fear?"

I feared everything.

My Lord and Savior Jesus Christ, whose tender voice filled

my every longing, now seemed as silent as the distant stars. I was left with my brain's revolt and low-slack collapse, a witness to my once reliable systems of adaptation buckling beneath the skin and the slow rise of a symptom I would come to know well.

Walking across Harvard Yard on a blustery night three weeks into my new reality, my legs suddenly forked and angled in a jumbled chaos, and I didn't know whether I'd make it back to my dorm. These legs, an abiding source of confidence, had earned me a coveted spot on the 440 relay as the fastest white boy in the ninth grade, holding the lead around the quarter stretch until delivering the baton to Willie Bell. Once reliable and sleek, these legs were now announcing to anyone paying attention under this low winter sky that, without strenuous conscious effort, they would not move forward toward Rockefeller Hall. There I stood, commanding them to move, the crowned king of freaks. Only after an hour of paralysis, of imagining I might have to crawl like a parodic pilgrim, did I find myself back in my narrow, windowed room.

I kept a copy of Oswald Chambers's *My Utmost for His Highest* up high in the closet, where other men might stash porn or a handgun. Next to the Holy Bible, evangelical Christians revered no book more than Chambers's devotional classic. Fellow believers cited the book with canonical reverence. My mother gave me the Dodd and Mead cloth edition the summer I began high school. It was a handsome volume, small enough to fit into the pocket of my backpack, and it had served me well as president of the Fellowship of Christian Athletes senior year of high school. I'd rarely opened it since, though it accompanied me each time I moved.

Now I turned to *Utmost* again mornings and evenings, hoping for a remedy or a healing word in its pounding insights. But something felt seriously off. The more I read, the worse I felt. Entry, Day 35: "If ever we are going to be made into wine, we will have to be crushed." Chambers demanded that I barrel through every compulsion inhibiting Christlikeness. I'd never considered what that involved. When you have a sound mind (and a white body), persecution tropes can feel inspiring.

I tried to muscle my way into Chambers's instructions for becoming Christlike. First, confess all your petty rebellions; God hates them. Next, confess your desires, even those you thought wholesome and true; repudiate them all, as Chambers does in painstaking detail in his 365 entries. Then gird yourself each day for battle; each day is a life-or-death struggle to be holy, a battering crusade against your wretchedness.

He sounded like a creepy youth pastor: "Be careful," "Be more careful," "Be careful to see," "Be careful to remember," "Be careful about the treasure," "Be careful to remain strenuous," "Be careful to keep the body undefiled," "Be careful to keep pace with God," and on and on. Depletion, ruin, your ash-heap life, the filth exposed—I am vile, God be praised. The kind of youth pastor who derives pleasure in ferreting out your intimate longings, who calls you on the phone if you miss Tuesday's discipleship group, who leaves notes in your school locker that he says are intended to edify but that make you feel like shit (because that's exactly how he wanted you to feel), and who (you one day realize) despises you as much as you despise yourself.

I had only to consider that my heart was hardened. I'd puffed myself up with knowledge and failed to honor the Almighty, and what I needed—it seemed clear—was to sink deeper into the Word.

What, then, does it mean to be a Christian? It means reckoning daily with your mind's exquisite corruptions, knowing you'll never reach their end. And you won't get any credit from God for honest self-inspection, for even the capacity to say "I am a nothing" suggests a sad withered leaf of agency—though agency nonetheless, that you dare not claim. Because without God you can't even know that you don't know anything. It means prostrating yourself before God, decreasing to zero, and (Chambers again) becoming his "devoted love slave." Anxiety, madness, the howling terrors, whatever powers are wrecking your mind—submit to them all for the sake of his pleasure. "There is no getting away from the penetration of Jesus."

Over the years I copied Chambers into my journal, and then echoes and twists from other reading, from sermons I heard, from Pope Gregory I to John Donne:

> *Count it all joy when you fall into trials and tribulations.*
> *Bear the scars of the Messiah, and rejoice.*
> *Consider how great a gift is bodily affliction, in that it both cleanses*
>   *and restrains.*
> *Each time a believer is chastised by God and becomes sick, he should*
>   *be glad.*
> *Do not scurry around in search of healing.*

Place yourself in submission to God.

Count it a privilege to suffer shame.

The father of spirits crushes us for our good, that we may share his
holiness.

Cursed is the crown; chastening the winds; the lack becomes the Lord.

Batter my heart, three-person'd God.

Here ends the lesson.

Autumn blurred into winter, and my waking hours gloomed in perpetual twilight. I kept my fears a secret, hidden in shame.

With each passing week, I was all the more sure that there was a malignancy in the world, and I was its source. Such became the mark of my high calling—surrounded by clashing armies, whispering sweet nothings for the gift of my nightmare life, to be open to God for the taking. I had no reason to hope that an armistice might have been reached the night before; any cease-fire would have been brokered only by alcohol and exhaustion.

December 23: "Remedy: (1) less coffee after a.m.; 1 cup + decaffeinated; (2) no tv at night (1 movie per week); (3) exercise: b-ball schedule; (4) more precise in daily goals; (5) meditation on the Word." And then on December 26: "What is the metaphor of people who break down?"

How I wish that young man could also have read Dorothee Soelle: "Affliction is regarded as human weakness that serves to demonstrate divine strength. Sickness and suffering are used for a religious purpose. . . . Corresponding to this tendency is the other, on the human side, to push for a willingness to suffer, which is called for as a universal Christian attitude. A person is denied the most elementary human right, namely, to defend himself and to say, like Goethe's wild rose, 'And I won't endure it.'"

———

Had you known me in those divinity school years, you might not have noticed anything unusual. In the tradition of the preacher's kid, I could strike a cynical pose. I kept a quiver of potshots that I could aim at the Moral Majority and the inerrantists. I could appear impatient with small talk and genially discontent with the state of things. I could lace my speech with expletives or smoke a joint. Aware of the suspicions surrounding evangelicals at Harvard and white Southerners in the North, I helped organize fundraisers for Food Not Bombs and the victims of the Greensboro Massacre (one of whom had graduated from HDS a few years earlier). The back hatch of my Honda Civic above the Georgia plates was a montage of bumper stickers meant to reassure the comrades: "No Nukes," "Carter for President," "Atomkraft? Nein Danke." (The last a palimpsest placed precisely to cover my once unapologetic Roll Tide.) To showcase my high-minded ideals on race relations, I told stories not entirely made up of my father staring down the White Knights of the Ku Klux Klan of Mississippi and of my own encounters with southern violence—once as a witness to a double homicide in a Pensacola bar—leaving the impression, I truly hoped, that I was worldly beyond my years and background. I had seen some shit.

It would have been easy to miss the way my eyes darted in fear of some new unsettling, or, if you had noticed, to not think much about it. I am unsure, now, how much even I thought about my state of mind—obsessively, some days, others not at all. It felt as though I'd been strapped into the command center of a mental radar system constantly surveilling the land and sky for trouble. Ping, ping, ping.

The only thing I knew for sure was the difference that November night had made.

"What does the Lord require but abasement," I wrote in my journal. I felt that I must discern my darkness, and therein remember that sin is not only what I did, but who I was: I am the one who breathes corruption.

How dark could it really get, you might wonder, in the mind of an evangelical virgin hoping to redeem the secular world? Pretty fucking dark.

The breakdown should not have come as a surprise. Impure thoughts had long ricocheted in my brain like bats trapped in an empty milk jug.

"You ascend the scale of erotic desire until you find God," a Roman Catholic friend once told me, a decade after I left Harvard. "Worldly phenomena are the glitter of the holy," he said. "All truth is God's truth." The two of us were sitting on a bench in the quad of the liberal arts college in Baltimore where we both taught. It was an afternoon in early May, a few days after classes ended, in a green trace outside the humanities building.

My colleague spoke with a gentle Irish lilt. He wrote ponderous books on selfhood—books constructed around the

---

I was moved by what my colleague said to me that day. It's a noble hope, the idea that our own erotic ascents lead us finally to God. And true. What my friend omitted, of course, was the postscript standardly attached in the tradition, viz., once you've arrived at that divine summit, your erotic pursuits retire into chastity. See Augustine; see Kenneth E. Kirk, *The Vision of God: The Christian Doctrine of the Summum Bonum* (London: Longmans, Green, 1932), 323–24.

peculiar language of existential phenomenology, but that included passages, in every chapter, that read as lyric. Philosophers with continental sensibilities (Hegel to Heidegger), as my colleague was, and theologians trained in modern German thought (Kant to Heidegger), as I had been, talked a lot in those years about (drumroll) "otherness" and "the other." It was "discourse" (ibid.) I found thrilling. This colleague had taken an interest in my work, and on that day he had a few comments on a conference paper I had written. But we both had a touch of spring fever, it would seem, as our minds drifted elsewhere.

I told him I wished I'd heard the gospel of the ascending erotic escalator in the churches of my childhood. I'd always assumed you could only find God by sailor-diving into guilt and shame.

"It's never too late," he said with a smile.

Oh, fair Liam, I do think it was.

———

On the afternoon before the breakdown, I'd been reading the assigned passages in a New Testament commentary written by a British Anglican teaching in the American Midwest. Stripping the gospel of all joy takes a certain gift, which this distinguished scholar possessed in spades. Each page opened to a spiritual wasteland. Gone was the poetry—gone the Swiss Alps and their votive morning mists, and the wands of flowers in the meadows nearby, where I'd read the Gospel of John during a summer at L'Abri, and gone the crickets' roar at summer camp, and the fabulous beat poetry of the testimonial. Gone was the heat rising off bodies in the hour of decision, the *amen*s and the praise. Gone was deep community; the mystery of second birth. Nietzsche would have loved it. I mean, really hated it, all the mollycoddling of the mundane.

"Behold, I make all things new." Read: "Try to stay calm and don't expect much." I'd been reading, specifically, in Luke 9, trying to understand it, as my professor thought I should, as a bloodless biographical apothegm. There's a demon-possessed boy in that chapter, but I didn't need to notice him yet, neither the way Jesus rebuked the demon nor the way the boy, healed, was sent back to his father.

In Cambridge, by the end of the semester, my hope for any healing seemed as long gone as the days when I would read Camus's "The Sea Close By" over and over on an empty trace on the Redneck Riviera, with a quart of Barq's root beer to salve the heat, sailing "across spaces so vast they seem unending."

I tried, a little obsessively, to appear an amiable member of the first-year class. I went to parties and concerts. I volunteered as an English tutor at a community center in the inner city. I kept track of my study hours—a minimum of six hours a day, seven days a week, recorded in my journal—and turned in papers on time. I took copious notes in class and typed them up in the evenings for my color-coded binders.

I paid attention to fashion as much as a student budget allowed, but no poplin and plaid for me. I steered toward seafaring collegiate with a twist of Neil Young, not quite *bonhomie academe*, but not quite anything really. My chronic skin allergies did not blend well at all with the nautical woolens of the era. The Nordic pullover welted my neck; my wool-lined wallabies caused constant and embarrassing foot sweat. In my brown corduroy jacket I was always either too cold or too hot. I studied in places I hoped would keep my symptoms at bay: beside a bank of windows in a church refectory or an empty classroom. I tried to keep a steady face.

The year wore on.

"You need Jesus Christ to give you strength in (1) purity, (2) dedication, (3) courage," my parents had written to me in a card on my fifteen birthday.

Of course I needed Jesus. I also needed professional help.

Could it have been more obvious? The mental upheavals of the midnight raid and so many convocations of dread, every lineament of my case, might have come straight from the *Diagnostic and Statistical Manual of Mental Disorders* (3rd edition, published one year earlier), of which I knew nothing. The DSM included in its description of Generalized Anxiety Disorder:* smothering phobias, hypersensitivity to noise, compulsive body monitoring, and the belief that you are going, or had already gone, insane. To which I could have added: eyelid twitches, random bestartlements, nocturnal anomie, muscle aches, and all the elaborate rituals to avoid the feared object.

But we did not do therapy—my family, my particular evangelical coterie. Os Guinness, the popular writer, speaker, and thinking man's evangelical, dismissed Freud and his pros-

---

* "The splitting of *DSM-II* anxiety neurosis into (i) GAD and (ii) PD in the DSM-III in 1980 was the official birth date of GAD as a diagnostic category."
—Marc-Antoine Crocq, *The History of Generalized Anxiety Disorder as a Diagnostic Category*

elytes and, really, all psychology: "The 'therapeutic movement' produces conceptions of the human that are 'all deceptive and all substitutes for God.'" Therapy loosens the fears and inhibitions that keep us unspotted from the world. The Christian should be grateful for repressed desires and a guilty conscience.

On a summer afternoon in Jackson, Mississippi, in the cool of my grandmother's dining room, the window unit humming like a kazoo, my cousin Bess had disclosed that she had started seeing a counselor. It didn't matter that my grandmother, the fragile daughter of a town drunk, had little love for my cousin's father and that she could always be counted on to provide safe harbor from his spare-the-rod-and-spoil-the-child fanaticism. Now she offered Bess a stern rebuke: "You best put an end to that right now, missy! All that's just going to mess you up more." My cousin did not say another word on the subject. Later, after college, she moved to Liberia to serve as a medical missionary in the aftermath of the 1980 coup.

That the past should not be questioned was Nana's first principle. Neither could it be redeemed. The past was a scar carved by providence into a sheet of granite; you could look upon in wonder, but you could not go through. Out of such brutal Calvinist realism, life's sorrows will be forever locked away in a private vault.

I wouldn't have therapy—not for some years anyway. And many days, it was not entirely clear whether—or, better, how—I had Jesus.

Dry Leaves Tumble Down University Circle

Dry leaves tumble down on University Circle. A tree outside the dining-room window stands guard over my privacy. A pop orange Bosch clock purchased on Etsy glares at me from the table. The timer functions for now, though the clock has stopped working, its hands hanging lifelessly at the six. My day proceeds in twenty-minute sprints. I break when the timer dings and rattles, get up, stretch, change the music, rearrange books. There was once a time when a student would, at least once each day, unlock the front door and tiptoe through the foyer and up the stairs to my wife's office—a volunteer helping stuff envelopes or make phone calls for K's nonprofit. But a year ago—or two or three—the nonprofit outgrew the third-floor garret and moved into a proper office. These days the house is quiet during work hours.

———

Charlottesville is full of apparitions but bereft of mystery. The town itself is as small as the towns I grew up in. Drive fifteen minutes from my house in any direction and you're surrounded by horse farms, rolling hills, and antique barns. Drive fifteen more, and you're traversing the lonesome upper hill country, with ole TJ himself preening over your academical village, where the southern nobility will come for reasons they do not fathom to drink the cup of knowledge, heading nowhere, like the world's biggest bullshit artists.

Until the summer of 2017, our cosmopolitan elites encountered mostly looks of unknowing when telling people outside the region or abroad where they came from. It took a short history lesson on colonial America to get an "Oh, yeah. I've heard of that," and only if you were lucky. But in August 2017, Charlottesville became #charlottesville, and suddenly you didn't have to say, "Not Charlotte. Charlottesville. It's where Thomas Jefferson lived. He had slaves." You no longer had to recycle the university's ridiculous boasting about Edgar Allan Poe (he lived in a room on the West Lawn for four months) and William Faulkner (he spent the least productive year of his adult life playing a fool on horseback and fox hunting drunk).

For the longest time, I found the town's geography baffling and could not orient myself to the four coordinates. I could not tell you whether the regional jets landing and taking off in the same flight path overhead were northbound to Philly and New York, southbound to Charlotte and Atlanta, or westbound to Chicago. Sunsets shadow the circumference of hills, and railways slither through ivied traces in patterns that confound the inner compass.

In my first home, I could hear the foghorns of barges and ships angling in and out of Mobile harbor. All the towns of my childhood lay near the Gulf, if not a short bike ride to the bay, an hour's drive. And after I left the South for college, the places I lived hugged the shore: in Wenham, Massachusetts, and then Cambridge, the Bronx, and then Baltimore. Oriented by water to the south or the east, I always knew where I was.

When we moved down to Virginia from Baltimore, K and I and our three children, it grieved me to think I'd lost the raveled city that had fed my soul and had come to feel like home. I was thrilled to land a job at a research university and even more to teach students whose backgrounds I better understood. (Who am I to cast aspersions, but I'd never found common ground with white Catholic Northeastern Beastie Boys wannabes.) But I missed the energies of urban life. I missed the half-hour commute to and from work and the fire and fury of local talk radio. I missed the dinner parties with our Mt. Washington neighbors, the bar mitzvahs and bat mitzvahs that our older son attended more often than confirmations—I'd not heard the word *mitzvah* until I was in my twenties.

Our first months in our new house on University Circle, I'd sometimes wake early on Sunday morning and drive the 162 miles to Baltimore's Episcopal Cathedral of the Incarnation in time for the eleven o'clock service. Once I locked my keys in the car, delaying the trip back to Charlottesville for as long as it took the Pop-a-Lock dispatcher in Dundalk to find a driver willing to work on a Ravens game day. I couldn't exactly explain why I returned to Maryland for Sunday worship—only gesture at what that cathedral held for me. All three children had been baptized there. There is a surfeit of material beauty there that captivated me during worship, so different from the stripped down and formal churches of my childhood. And there was a tie-dyed woman who every Sunday stood up during the Prayers of the People and prayed for Jerry Garcia—that was a blessing I didn't see coming.

Sometime after joining the cathedral, I was given the keys to an unused classroom, light-filled and right next to the pipe organ. On writing days, I'd taken to attending morning prayer, and sometimes at odd hours I sat in the quiet of the sanctuary, captive to a silence I was only learning to call prayer. The cathedral, then, was not only a place I prayed, but also a place where I wrote—and during those hundreds of hours of reading and writing about the Mississippi civil rights movement and the Christians who fomented and opposed it, I began to take apart and put together stories that, I was realizing, had been the scaffolding of my childhood, of my own education in those Mississippi schools and in my father's church. One weekday morning, I sat in the sanctuary not certain I could write the book after all, and there's no way to say what happened other than to tell you directly that Fannie Lou Hamer herself, then dead fifteen years, came to that sanctuary and comforted me.

In Charlottesville, nearly a year passed before I met our next-door neighbor. Around the same time, the old-money divorcee in the self-satisfied Tudor paid a visit, only to demand that our children stop cutting through her yard on their way to school. They were using a foot path on the border of her property to avoid a curve on busy Rugby Road. For a long time it felt as if all the bright land of Albemarle County held life in like a deep breath.

Charlottesville was not entirely unknown, of course, when K and I moved here in the first summer of the new millennium. And that was its own strangeness, to be back as a member of the faculty I had fairly despised as a grad student a decade earlier. The white scholar of indigenous American religions who once told me I could not borrow his lawn mower because it was the "Mercedes Benz of lawn mowers." A Jansenist who'd refused to intervene when three women students complained of being sexually harassed by another Jansenist. The philosophical theologian who walked out of the seminar room in a quiet rage when questions were raised about Heidegger's Nazism.

Why did I move back here? Was it only to resume psychotherapy with the woman I imagined had saved my life when I was twenty-nine?

Hired with tenure after teaching ten years at the least known of the four Loyolas, I felt, it must be said, an unworthiness to be joining one of the best religious studies departments in the country. One of my new colleagues—a bald battering ram of eloquence forged in a famous Northeastern boarding school— told me point blank, "Charles, you are redolent of imposter syndrome." His bluster took me by surprise. I'd worked hard in the decade since doctoral studies, written books, published technical papers in peer-reviewed journals, won awards. All of which I mentioned when I told him he was wrong. I did not have imposter syndrome.

I had an anxiety disorder, which of course I kept to myself. Early on in my first semester back at UVA, I noted in class that

Marx had said the purpose of life was not to understand the world, but to change it. And immediately I knew I was wrong. Not only had I botched the phrase, but I cited the source as the *Communist Manifesto*. It felt like the time I attended a debutante party in Atlanta and told a girl and her escort that "I were glad to meet you" (the exact words that rolled off my tongue). I'd spent a month reading Marx with a research group at Fordham University and a sabbatical semester in Heidelberg studying Hegel's phenomenology (where Marx found his dialectic, and where I found anxiety triggers). I tried to pause, but it was too late for retractions, and I couldn't collect myself. A goateed PhD student informed me with a grim smile that the quotation could be found in Marx's *Theses on Feuerbach*, the eleventh thesis to be precise, and that the correct rendering was, "The philosophers have only interpreted the world in various ways; the point is to change it."

Later that semester, or maybe it was a different semester, I found myself lecturing from typed notes on a topic that in those years I could usually talk about with some competence. How, according to theologian M, let's say, God is essentially *act*, while according to theologian P—archnemesis of M—the idea of God is better identified with *being*. It's the whole transcendent versus immanent debate, and you're in good company if your eyes are beginning to glaze over. And yet they were lines in the sand for the men and women—mostly men—who see pure doctrine on one side and heresy on the other.

But then, with the shock of a man who unwittingly hollers out a word or phrase of an interior monologue, I suddenly could not remember why the distinction mattered. Oh, if you stopped by for office hours, I could reach splendid heights on the story of modern theology as it came down to a standoff between the Swiss German Karl Barth and the German German Friedrich Schleiermacher. But with the midafternoon light bending toward sunset and the campus ablaze in serotine glory, I couldn't remember a single thing about the distrust between the liberals and the traditionalists or what was at stake in whether one thinks of God on the basis of God's transcendence or on the basis of God's immanence. The notes that had seemed so lucid the night before—that I had worked on until the early hours, that I had, let me confess, cut and pasted from my doctoral dissertation, and that had included an invocation of an article by my *Doktorvater* (that's the German term for mentor/tormentor) entitled "The Being of God When God Is Not Being God"— looked as strange as hieroglyphics.

I could have called a quick break. Instead, I kept talking. I don't know what I said, only that I lumbered onward, skipping entire paragraphs in my lecture notes to land randomly on a passage that made no sense without context, if it ever indeed made sense.

Then suddenly, in the midst of that professorial debacle, an old memory nearly forgotten appeared: a childhood trip to Montgomery and a man in a white coat telling my fragile mother that her eight-year-old son needed a surgical tongue trim. And a diagram of an open human head and plum-red palate as thick as a Sunday roast. "Let's slice it by 20 percent. You'll be in and out in an hour."

My dear mother, you took my hand and led me back to

safety. To an afternoon matinee of *The Sound of Music* and then the highway home under the sunset sky. *But hey, Mom, WTF?*

———

November 2019: I wake at first light. Virginia creeper in serious need of trimming canopies the bedroom windows in soft autumnal violet. The line comes to mind: "His compassions are new every morning." I read in a copy of F. B. Meyer given me by my grandmother that union with Christ is like "the Amazon River flowing down to water a single daisy."

After coffee, I open my computer to a recent Gallup survey on religion and mental health. A research team at Baylor University studied 1,714 respondents who were asked such questions as "Over the past month, how often have you felt nervous, anxious, or on edge?" and "Does God help you feel better?" The subjects admitted to having prayed somewhere between "several times per week" and "once a day" in the preceding years. The researchers concluded that "prayer doesn't ease symptoms of anxiety-related disorders for everyone." Indeed, some feel that prayers are mostly desperate attempts to keep their shit together.

I must admit I find this not surprising at all.

Later that day, a box arrives in the mail with an Oregon return address. It contains a blue medicine bag so small it would not hold a vial of pills. This is one of the many items I've recently purchased online that I don't need. Until just now, I'd forgotten about it completely.

Other such purchases include:

- A cone-shaped hat made by a Japanese designer in New York, which, my dining-room mirror suggests, casts me as an effete Klansman.
- A half dozen artisanally crafted tote bags from a boutique run by two friendly lesbians in Arkansas (which would be a tremendous help in my daily tasks and a thrill to sport if I had seven arms).
- A versicolor print of the human heart featuring three comic-book heroes launched from the red beating organ to save the day.
- A burlap work shirt that the bespoke tailor called "keen classic chore" but that sent fiery whelps down my neck, calmed only by cortisone cream.
- A gun-metal green storage box too narrow for CDs and too wide for business cards (which I hardly used), with a rusty bezel exactly planed against an interior slat that, the day I opened it, sliced my index finger and left a trail of blood.

Still, in recent months (my wife would say years), the written word sings to me of everlasting days, and I'm swayed by its lure and charm. Which is to say, I've lost control over book buying. I know of what Eudora Welty spoke when she said, "Of course I'll not read all of these books." But Eudora— who lived a hop, a skip, and a jump from my grandmother (Nan Toler: 917 Fairview; Eudora: 1119 Pinehurst); the two were friendly when they ran into each other at the Jitney

Jungle, though Nana didn't like her stories much and attributed her spinsterhood to bad posture and a weak chin—never had to deal with one-stop online shopping. I've clicked "Order Now" on treadmills, airplanes, and commodes, in bathtubs and barber shops, while sitting in my car at a red light and walking the dog in the afternoons. The rush of the click and the beep of the receipt immediately transported to my inbox ignite the hope that I will have time to devour them all.

I'd once lost a girlfriend in New England over books, a tall blonde Marxist evangelical named Maggs from Southern California who'd earlier in the semester surprised me with a hand job on Singing Beach—my first at Singing Beach, or anywhere—and would have been happy to go all the way. We were sitting on the steps of the college library. She said she enjoyed my company and wished things had turned out differently, but she could not compete with my mistress and was tired of trying; she needed to move on with her life. She was thinking of me and books.

Indeed, I loved words; the more floral, the more I loved them. Words might not always be safe or comforting, but words were familiar. Books didn't make me feel guilty. I could arrange books on a shelf, view them with detachment or interest, touch them or leave them alone.

Do the math on this. It took a week and a half to get through a three-hundred-page thriller set in the rural South. What's my plan for Osterhammel's thousand-page *The Transformation of the World: A Global History of the Nineteenth Century*?

I don't have a plan, but I do hope in the resurrection of the dead. I imagine heaven as a place of leisurely reading interrupted by pickup basketball games in which—thanks to my resurrected body—I'll run without pain, regain my long-lost defensive skills, and maybe even dunk on occasion. Of course, to hope is not to know.

———

I didn't realize it until long after they'd been purchased—plucked from a corner shelf at a shop on Elliewood Avenue, scooped up from the front table at Barnes and Noble, scavenged at the annual Friends of the Public Library sale—but in those early years back in Charlottesville I bought a great many books on the theme of nervous breakdowns—their etiologies, their choreographies, and how to survive them (or not) with flair. *The Bell Jar, Lancelot, Tender Is the Night, The Awakening, Notes from Underground, Briefing for a Descent into Hell.* What strikes me now, in the company of Miss Eudora's insight, is that the books on breakdowns tend to be among those volumes on these shelves that I have read.

In a monograph I perused one afternoon, I found the observation that in fin de siècle fiction the "new woman" was prone to nervous disorders—the novelist's point being that her ideals were too advanced for her hidebound culture. It's no surprise that I was attracted to such an argument, which twined into one fine floss my own sophistication and a critique of the clutches of Christian choreographies I simultaneously relied upon and deplored.

Still, the novels and histories of madness couldn't hold a candle—well, maybe Plath could—to stories of the Complete

Nervous Breakdown I'd heard throughout childhood. My grand-mother always had a story about somebody she knew who'd broken down. One involved a niece whose husband proved im-potent in the marital bed. I think it was the man in my grand-mother's story who cracked under pressure, but it may have been the woman. The woman, my grandmother said, had moved to Arizona and joined a cult. The man attacked the family attor-ney with a hatchet and then went home, fixed himself a cocktail, donned one of his wife's dresses, and took a seat on the porch to await the police. (So I'm guessing it was the man who cracked.) Other stories followed the short tragic drive from Jackson to Whitfield, where a friend or friend of a friend had been com-mitted in the state lunatic asylum, after which, everybody knew, nothing would ever be the same.

My mother talked a lot about nervous breakdowns too. While Nana would as soon tell her story over a plate of fried oysters at the Mayflower Café as in the privacy of her own kitchen, mother spoke in hushed tones of white people gone mad and swore me to silence. She was partial to tales of ladies with refined tastes snapping under pressure.

I had stories of my own. My piano teacher's husband—once a prominent lawyer—lived in the back room of their home near the city cemetery. Each week I shambled through the same Bach minuet, getting nowhere, while the poor man banged around in his quarters. It sounded like a moving crew in there. Sometimes he would appear in the hallway of the music room, take a polite bow, and silently clap his hands. I silently shrieked in terror inside from head to toe.

You had to be careful. A line ran from the bedroom to the

madhouse through the parked car and the cheap hotel, from hanky-panky to the CNB (Complete Nervous Breakdown). Both my maternal intimates agreed on this—and neither spared graphic details.

Now it strikes me that the shape of these stories is, while unmistakably Southern, essentially, Christian, the parallel lines of the complete breakdown and my total depravity merging finally into the one either/or: redeemed or damned; pure or impure; all-in or lukewarm; possessed of the mind of Christ or strapped into a white paneled van on the way to Whitfield. A word could heal or wound, edify or shred. I never knew where I stood with the one or the other. Fear was the only constant. Too much depended on the right word spoken.

———

An engine guns at the corner of the street, screeches through the sharp curve, and passes my house on the wrong side of the one-way street. It happens about once a month that a student behind the wheel of an enormous SUV will turn onto University Circle at the intersection near Lambeth Field, ignoring the one-way signs. How there haven't been more serious accidents—only a few in the twenty years I've lived on the street and no fatalities—is one of the mysteries of college life. Another being why students are not falling off rooftops and poorly constructed balconies on the hour. The weather is warm and getting warmer.

I walk onto the front porch for a look, as the car ascends the hill back to Rugby Road without a head-on collision or killing a pedestrian. For the moment, I hear only the roar of cicadas, rattling their tiny tymbals like drums in the aging pines.

I offer a prayer of gratitude. It's become my only prayer, most often whispered as a benediction of duty, free-falling late at night into the ethically sourced, soft-jersey oblivion of rest: "Thank you, dear Lord." Is this how it feels to die a happy death?

The air is soft and warm. This is the best hour of the day. The hour of concentrated stillness. When things done and mostly left undone reach a juicy ripeness. The westering sky burns orange beyond the tree line. If I were standing in a field, I would raise my hands to the sky and welcome the night.

PART II

My Rebel Flesh

On Fire

It's nine o'clock on a Sunday morning, around the time my Uncle Scott gave me Sly and the Family Stone's *Greatest Hits* for Christmas. It was sometime between the bombing of Cambodia and the Watergate trial. As ever, I sit in the balcony at church. My father, the pastor, is in the pulpit. I am in the throes of arousal. My roving eye lands on the deep buttery thighs of Parker Ainsworth, who sits upright, and the pink trim of her panties, thanks to the recent popularity of the miniskirt. The affluence of Laurel—of the white professionals and the titans of timber and oil, at any rate—enabled a haute couture unrivaled by most southern towns. Next to Parker sit other girls my age—Jaynie and June, Debbie and Shauna, Delilah and Jezebel—with crossed legs, supple, tanned, and silky.*

I look away, bow my head in prayer, and search my head for Bible tips. "Whatsoever things are true, whatsoever things are honest, whatsoever things are just, whatsoever things are pure,

---

* I'd once asked the girls in my youth group to dress more modestly, in accordance with 1 Timothy and out of concern for a young's man struggles, but the admonition fell on deaf ears.

etc., think on these things." I think of Jesus's bleeding, tortured, graphic death walk to Golgotha, a walk made for sinners like me. Embrace the torment, young man. But the shift and twinge in my lap begs for attention and a subtle rearranging.

She fluffs her skirt, shifts in her chair, and crosses her legs, thanks be to God. Though not enough, it would seem, to veil the flesh: her panties form a wedge between her thighs the color of ripe peaches. I take advantage of a heads-bowed, eyes-closed prayer to steal a glance, knowing full well that Jesus is coming again soon, and it doesn't look good for me. The rough, sad sounds of everything east of Eden whiffle in my brain.

How hard it is to bring erections under the lordship of Christ. Not even my improvised chastity belt could help: tighty-whities secured by a jock strap so tight it squeezed my junk like some kind of BDSM zentai. Sometimes I'd stretch my penis toward my abductor magnus as if trying to engage it in the dank chamber of my upper thighs bulldog-style. In time, I would find it helpful after temptation and fall to rub a rough dish towel against my genitalia until I bled.

———

I was delivered into the world by a man forever known in the Marsh family as "that brilliant Jewish doctor" (in my parents' mind, the "Chosen People" occupied a place reserved for people of superior wisdom—though they had no Jewish friends, so far as I knew). Though my parents hoped for more children, more children never came. Where my mother tucked her sorrows over this I'll never fully know. She said to me, often and regularly, that she wanted a houseful of children, but that God blessed her with only me, her special one, who could carry the blessing of all the children she deserved.

Always left open—till the day I left for college, when my mother was thirty-nine—was the prospect that she might yet get pregnant again. At first, the promise of a brother or sister was welcome, so welcome that I created an imaginary brother for myself, Michael, who kept me company those years when we lived out in the country. But in time I developed the sense that I did not want brothers or sisters, that any sibling would ruin the heightened intimacy I felt with my parents, and especially with my mother, and at that point Michael disappeared. He was replaced by a feeling of barely articulable unsettlement— the uneasy concern that something would undo our shocking familial closeness or, worse, the concern that nothing would undo it.

And so it was a small and close family of three that moved from south Alabama to Laurel, Mississippi, in the late summer of 1967. My father was the new pastor of the First Baptist Church. In the family photograph in the *Laurel Leader Call* announcing our arrival, I wear a plaid sports jacket, a white dress shirt with a clip-on tie, and gray gabardine slacks riding high on my penny loafers. My pretty mother stands next to me with perfect posture, one hand resting on my shoulder. The tips of my shoes touch in a kiss. And my father—six foot two, athletic, and tanned—my God, is he handsome.

In the first month, I found fishhooks in two hamburgers ordered at different diners. A pack of boys shrieking like the *Straw Dogs* rapists led me to a remote corner of the parking lot and took turns farting in my face. How that was biologically possible was a thought that would only later cross my mind. On our first Sunday, a man who the very night before had brought us a large tin container of smoked sausage climbed a ladder to fetch his daughter's cat from the roof of their house, slipped, and died. On my first day of school, a Catholic kid named Reilly, son of the local golf pro, curled his fist during recess and punched me in the crotch for no reason I could fathom, except that, you know, sinners like me deserved a random shot to the nuts.

Everyone was angry and aggressive, even the church ushers. It is true and widely observed that there was no hour of the week more segregated than eleven o'clock Sunday morning. Less known is that no hour was more dreaded by an anxious twelve-year-old preacher's kid recently aware of four congenitally missing front teeth. It didn't matter where I sat in the cavernous sanctuary; a wiry bachelor named (let's say) Harry Pointer would hunt me down and aim his prickly knuckles at my thigh. "A charley horse for Charlie," he would say, as I smiled politely, wanting to kill him.

One afternoon shortly after our move, I lay on the bedroom floor poring over maps of American towns and cities, envying the geometry of urban grids and rural greens—a pastime that baffled my parents but felt reassuring to me—when I felt a sudden constriction in my chest. I sent out a raspy SOS to my mother in the kitchen, who found me half naked on tiptoes, pounding my chest like a wild child. "Breathing!" I managed. "Not breathing!"

She did her best to help, treating my wheezing and panic

with steam from the shower and a wet cloth on my neck. Of course, it was not the last time asthma would visit.

I'd never had allergies before, but now I was allergic to everything. Dust, grasses, feathers, hay, plastic, perfumes, and wool—and on top of all this, Laurel was home to the largest synthetic wood manufacturer in the world, which produced a pungent orange haze of God-knows-what that Stevie Forbert in his rock star years memorialized in lyrics: "Going to Laurel, it's a dirty stinking town, yeah."

Each day brought a new catalog of welts and hives and itches and runaway fluids. My nose was a constant phlegm leak. When the steam or VapoRub didn't help, my mother took me to the doctor for an injection of something that felt great but kept me up most of the night. I was kept awake too by an absolute terror tunneling through the mind and body—of never again being able to breathe.

"An asthma attack feels like two walls drawn closer and closer, until they are pressed together," wrote John Updike. "Your back begins to hurt, between the shoulder blades, and you hunch. I could not stand up straight and looked down at the flourishing grass between the sandstones. I thought, *This is the last thing I'll see. This is death.*"

———

And then a ray of light appeared my first year in Laurel like the star that rose over Bethlehem: I was tapped to serve as the narrator in the annual Christmas pageant. It was an honor to be asked and the first time I assumed a public role performing as the minister's son—well, other than the fact that my whole life had been a performance as a minister's son—and I took it very seriously. It was an opportunity to shine at an awkward age, to present myself as having my own talents and capabilities. Also, it got me out of Sunday evening services; rehearsals were then.

Throughout the weeks of rehearsals, Harlan Bush, aka "Flame," the local fire chief who volunteered his services each year as the pageant's director, seemed reasonably satisfied with my progress. But two things went amiss that December night in 1970 in the public auditorium at Gardner Park. The first is that I began to feel extremely uncomfortable in my slacks, which were of course too tight—I'd grown six inches between sixth and seventh grade. I wasn't aware of my response to the constriction, focused as I was on getting the baby Jesus born, until my father's carefully worded caution during intermission— "Son, you may not be aware of it, but you've been scratching your wikky a lot"—which made sense of the gaggle of high-school girls giggling in the front row of the audience.

So I retreated to the bathroom to make the proper adjustments and returned to my marquee spot at the podium ready to shine brightly as the northern star that guided the holy family to safety. But when the lights came on and the curtains were drawn, the second really bad thing of the evening happened. I was standing on a three-foot-high platform with an eight-piece brass ensemble to my left, the pianist to my right, and the choir behind me. I was to speak my lines—my earnestly memorized lines—into a squat mike on a stand. Leaning in to annunciate the Magi's arrival, I leaned too far, and my forehead made contact with the microphone. It was truly horrible to behold: the microphone and stand slowly fell to the floor in the silent auditorium, and although most everyone saw what was happening, no one was prepared for the riot of a wired mike hitting the tiled floor. An errant crop duster missiling into the mezzanine would have made less noise.

I became aware that Mr. Bush was approaching from stage right. A thick, damp hand grasped my clip-on tie, and, in a whisper loud enough for everyone in the vicinity to hear, he said "You have single-handedly destroyed the story of Christmas."

———

I didn't understand, of course, that we had moved in the fall of 1967 to a town that the FBI considered the epicenter of southern white terrorism. Yes, *Brown v Board of Education* had been the law of the land longer than I'd been alive, and yes, LBJ had signed the Civil Rights Act and the Voting Rights Act, but in deepest Mississippi the explicit agonies of the civil rights struggle were still unfolding. In the fall of 1967, in the courthouse in Meridian, federal prosecutors tried eighteen white Mississippians for their role in murdering civil rights workers Michael Schwerner, James Chaney, and Andrew Goodman.

Laurel's own Sam Bowers would be one of the seven found guilty in the courtroom of William Harold Cox (a Kennedy appointee who, despite once opining, in a 1962 decision about voting rights, that "the intelligence of the colored people don't [*sic*] compare ratio-wise to white people," and despite once having been described as "a master of obstruction, and delay," quite possibly "the greatest single obstacle to equal justice in the South," and despite holding views that, in the estimation of political scientist Charles V. Hamilton, "were directly contradictory to . . . the Fourteenth and Fifteenth Amendments," did insist that he would not "allow a farce to be made of this trial").

In the fall of 1967, the Klan bombed Beth Israel synagogue in Jackson, and two months later the home of Beth Israel's rabbi. In the fall of 1967, the Laurel home of a prominent Black minister and NAACP leader who'd dared, in the words of the local paper, to work "with white leaders to promote harmony between the races," was likewise bombed, as were the offices of the *Laurel Leader Call*—not because the local newspaper had said something against the Klan, but because it had not said anything in support of it. (Not for nothing did a respected journalist title his recent book *When Evil Lived in Laurel*.)

Thus the setting of my first year at that feral palaestra Mason Primary, an oblong mass of block and steel that could have passed as a minimum-security prison long overdue for basic maintenance. Is the dread I still remember when I think about my school days in Laurel merely the ordinary dread that smart children in schools that shun academic aspiration always feel? That was, to be sure, ingredient.

But for children in Mississippi in 1967 there was something more. The all-pervading terror of white racist violence, and the ways "Negro schools" both were and weren't a refuge from that terror, had long shaped the schooldays of Black children. In the 1960s, white children's schooldays came to know a fainter, but still spiky version of analogous dread. Configuring my mornings at Mason Primary was the sense, conveyed by most of the school's teachers and staff, that the school itself was the site of a great cosmic struggle: the public schools might be forced to racially integrate, and if this happened, all would be lost—our purity, our security, our way of life. One day, the principal came on the loudspeaker to announce that Martin Luther King Jr. had been shot and killed in Memphis. Our teacher's mouth admitted a rare smile like a crack in a porcelain bowl, and my classmates erupted in cheers. (Often, when I have remembered that moment, I have paused, because I find it so incredible. I pause to say to myself, *That couldn't really have happened, Charles,* only to remember with incredible vividness that my classmates did erupt in cheers, and I may have too.)

———

Our house sat near the end of a cul-de-sac in a subdivision surrounded by woodlands. To get to the house you had to turn off a lightly trafficked highway and navigate a labyrinth of interlacing streets until our street, Highland Woods, appeared like a tunnel cut into alluvial darkness. Wild sugarcane as high as a basketball goal and too thick to penetrate abutted our backyard like a border wall. Floodlights swept the yard after nightfall; and for added measure, a cowbell was affixed to the front and back doors to alert us to trespassers not already sniffed out and attacked by our beagle and the other dogs who roamed the neighborhood at all hours.

I don't recall ever opening a window in the house. Not even on spring afternoons when the yard swam in rose-colored light and honeysuckle. From the look of it, I don't think anyone in the house's thirty-eight years had ever opened a window. All had long been encased in burglar bars and painted shut. If you wanted fresh air, you could go outside to get it.

My parents never owned firearms and said no every time I begged for a shotgun. But they kept hammers, a souvenir tomahawk, and blunt objects in easy-to-find places. On the occasion when my father left town to preach a revival, my mother and I brought the weapons with us to bed.

Any stranger hoping to make our street a shortcut to 6th Avenue could expect a visit from the dogs patrolling the tiered lawns unleashed. Our dog, a wiry beagle named Spot, lived mainly in the wild and at least twice a year snipped or full-on bit the postman or a dark-suited missionary from an off-brand sect—and no one got too upset about it; we were in this together.

You couldn't have found a safer neighborhood in those years; no one ever recalled a break-in. Yet we lived, mother, father, and only child, with the certainty that we'd be invaded. Communists, Leninists, Stalinists, unions, beatniks, peace groups, SCLC, SNCC, COFO, CORE, GROW, NAACP, Northern clergy, ecumenical counsels, globalists of all sorts— they had us in their sights.

No God-fearing Baptist I knew read the dispatches of the White Citizens' Councils as fiction. Tribes of Black Bantus cowered in the lowlands near the Nam Cam Projects, surviving on crawdads and turnip root; they had weapons stockpiled and ready for a scorched-earth crusade to destroy the last vestiges of Anglo-Saxon Christianity. And to their rear, an Afro-American government, hitherto exiled in Cuba, was ready to storm the coast with Soviet-supplied implements of destruction. Come, Lord Jesus. It was that crazy.

When in my early thirties I revisited these years as a scholar, I was struck by how our siege mentality had become so pervasive, so complete as to constitute an epistemic foundation upon which rested our entire worldview. We fashioned our own lingua franca in which there was inner coherence, but a breakdown of ordinary meaning—*holiness* came to mean *power*; *civilized* came to mean *violent*. At the time, however, I was aware mostly of principalities and powers colliding and clashing in my body. If I had to be more specific, I'd say they entered the diaphragm and abdominal organs and rubbed gut against tissue and bile until smoking vapors ascended into the head like angry clouds. Or maybe I'd say this: we were convinced of our own superior righteousness and of the threats mounting daily and everywhere against it; no people in history had ever been persecuted like white segregationist Christians.

Alone at night, with the bedroom shades drawn and burglar bars secured, I knew no other way to calm my fears than to call out to my parents for help. My father might appear with an exhortation to count sheep and a perfunctory pat on the back. He ascribed to the theory of mind over matter when it came to his son's nocturnal terrors. Sometimes it worked; I might count down from fifty or imagine a comforting scene piece by piece and fall asleep. But mostly his counsel made me more agitated.

It was my mother's body that was the body of consolation, and most nights I would not give up until I got what I wanted—my mother's presence in bed. I wanted her to come to my room and lie down beside me. Was she wrong to oblige me, time and again? The manuals called for cold turkey. Let him sweat it out. He'll be fine in a week or two. If it gets bad, try rational reassurance and a loving rub. Point out the burglar bars, floodlights, and dogs. Close with a prayer.

In the mornings, I woke up alone and swiped away thoughts of the night. It was time to propel myself again into a new day like any pubescent boy in his stiff Levi's trying to forget he'd slept with mom the night before. After school and sports practices ended, with supper plates cleaned away, bedtime loomed again, when the room grew dark and my body jumped at every sound; there was no silence, just the dread of another tug of war that this time I might not win.

What do you do with a frightened child?

———

Jesus was set to return soon, and very soon. Despite my disciplines of self-denial and mental fitness, I believed in my heart of hearts I wouldn't make the cut on judgment day. I'd found nude photos torn from a magazine in the woods one afternoon—the detritus of another doomed Baptist boy—and had not disposed of them properly; by which I mean I had not set them on fire as I had *The Chipmunks Sing the Beatles* and, more recently, my *Easy*

Rider poster and a Ouija board. I had not plucked out my roving eye, which I should have done, because, you know, as the good Lord himself said, it would be better to enter the kingdom of God with one eye than go to hell with two. And I was pretty sure Jesus would find me humped over a rotting log huffing nudies when the trumpets blasted and the sun went dark.

I knew my body was the temple of the Holy Spirit. The goal was to remain sexually pure—presenting the temple undefiled—until I met my soul mate, who had made the same arduous journey. My family, friends, church, and the whole *sanctorum communio* prayed that I would run and finish the race; and they prayed for my future mate as well, for the one who had been chosen as my helpmate before the creation of the world, an idea no less grand in God's perfect providence than the firmament of the heavens and the light upon the earth. I had only to wait to receive my just deserts.

"You need Jesus Christ to give you strength in (1) purity, (2) dedication, (3) courage," my parents had written in that birthday letter. My mother explained to me that premarital sex leads to psychic ruin. "All the girls I know who've lost their purity have emotional scars. They've lost a precious something they can never get back. Their thinking is somehow damaged."

Those girls had committed the unpardonable sin, was my takeaway. *Oh, Lord have mercy, the unpardonable sin!* From my first reading of the Gospel of Matthew—"Wherefore I say unto you, 'All manner of sin and blasphemy shall be forgiven unto men: but the blasphemy against the Holy Ghost shall not be forgiven unto men'"—I was consumed by the need to understand this. Here was Jesus, turn-the-other-cheek, blessed-are-the-merciful Jesus, showing us the dark hole in his big love. All sins would be forgiven but one. And he wasn't saying which. But I was pretty sure I knew.

The problem was that the more I learned about my contemptuous flesh, the more I wanted to feel the shape and heat of my girlfriend's lap, where she would let my hand rest sometimes at the movies. She was probably not my special someone, nor I hers. I was fairly certain that Jesus would return to rapture the Christians before I could marry and go the distance. The lines of desire and purity collided in a heap of shame.

Numerous books on the subject of a young man's struggles appeared mysteriously in my bedroom during these years, sometimes inscribed by a parent, other times randomly tucked into my bookshelf. They reinforced the life expected of me:

> The most wonderful gift you can give your future bride or groom is your purity.
>
> Strip tease and leg show, bathing suits which unduly expose the body, particularly women's bodies, magazine stories and pictures that turn the mind especially toward love-making, the movies, the embrace of the dance—these lead to sex desire and so to necking and petting.
>
> God wants more than just a person's mind or his service in some future career. He wants his body kept pure and clean. Now!

In my journal, I charted the slippery slope of the affections: "1. Acquaintance. 2. Casual Friendship. 3. Close Friendship. 4. Intiment [*sic*] Friendship. 5. Immorality/Impurity." I fretted over the ease with which I could become, as I wrote, "Satan's main instrument."

So I equipped myself for battle like St. Anthony in the desert. I would fight the good fight. And I would lose my mind anyway.

———

My descent into juvenile delinquency began as a spontaneous act as seemingly unmotivated as Meursault's killing the Arab. I was walking on the sidewalk one afternoon along Bay Springs Road drinking a bottle of pop with my best friend, Mike West. I could see the motorcycle approaching from the north near the curb store, but didn't think much about it until the Harley downshifted a few car lengths away. When the bike reached us on the road, I spun around and sidearmed the bottle straight at the driver's head. "What the hell?" Mike shouted. "What the burning hell?" We jumped over a copse and tore through the kudzu screen into a jungly escape.

Pubescence came over me white-hot and fuming. The way I saw it, as long as I remained pure—didn't have intercourse before marriage and abstained from drugs and alcohol—I was free to run wild. So I turned to throwing bottles and water balloons at passing vehicles. I punked M-80s into the exhaust pipes of parked cars—the kind in which couples drank and fornicated—and watched from a safe distance as the tailpipes convulsed and flared like fabulous roman candles. I fired bottle rockets from the bridges and overlooks of the interstate and outran cops and highway patrolmen, because I knew the trails that cut through alligator weed and honeysuckle back to Ellisville Road.

My father, bless his heart, and all of our parents were clueless. My father would lead the weekly meetings of the Royal Ambassadors, awesome gatherings that included a Bible study and film highlights of SEC football games, and then dismiss into the night a pack of hooligans with red-letter Bibles hooked on bedlam and the destruction of property.

On Wednesday nights, my friends and I had taken to bailing on choir rehearsal, instead playing basketball in the corner lot. We no longer wanted to sing in the music minister's sappy Christian musicals—we'd moved on to Sly and the Family Stone and ZZ Top. One week, we found the church basketball hoops removed. In retaliation, Derek Ham and I set the music minister's house on fire—though arson was not our intent. Camouflaged by purple pyramids of sweet gum, we bombarded the house with bottle rockets—just intending noise, just intending to disturb the despised minister's sleep. Still, I wasn't unhappy when a rocket landed on a bed of pine needles on the roof of the house, smoldered, and then burst into flames.

If you're a fundamentalist boy coming of age at the exact spot where the Bible belt breaks the skin—preinternet, pre-Gameboy, premillennial, pre-everything—and there's nothing to do afternoons and weekends but loiter outside the curb store, you've probably already discovered the thrill of juvenile delinquency.

The fire has to go somewhere.

The Pursuit
of a Literary Life

It's common knowledge that in Mississippi writers emerge from hamlets and hollers like the unstoppable creep of kudzu vine. Ralph Eubanks, in his recent literary history, explained it as the result of historical dissonance and perpetual civil unrest:

> Like Ireland's, Mississippi's history is filled with suffering that must be explained; it is a place that comes alive in its stories and inspires those stories, which flow through every bend of its winding rivers and across every piece of land within its borders. It is the beauty of the land mixed with the state's complex history that inspires and perplexes its writers.

Or as Richard Ford, the novelist who shares with Eubanks a devotion, in particular, to the short stories of Eudora Welty, says: "When you have built into your society a completely irreconcilable human conflict, slavery and segregation for instance, there are schisms and torques and breakage all around you, both about race and not about race, drama in other words."

All the late greats—from William Faulkner and Richard Wright to Eudora Welty, Elizabeth Spencer, and Walker Percy—had their own take on why this weird, wan place that nobody wants to visit produced our most arresting canon. But it's a mystery to me how those greats came by their artistic formation. Because I don't remember ever reading a work of literature—beyond the Bible, that is—in the six years I lived there, or ever being asked to.

Entire school years passed in a blur of unknowing. From the fourth through the sixth grade, "The Lottery" and "Silent Snow, Secret Snow" wave in faint recognition. Not a single book was assigned in junior high, though I tried to read one of James Oliver Kerwood's dog novels on my own.

It's many a writer's boast to enumerate the heavyweights devoured while the rest of the world frolicked in the nettles. I was not a precocious reader as a child. You will never find me among those connoisseurs of writerly writing in the Sunday *Times* who, week after week, wow all of us sleepy, hungover imposters with astonishing feats of early reading: English literature from Austen to Yates by sixteen. Which is not to say I did not read at all—in addition to trying Kerwood (*Kazan? Son of Kazan?* I can't remember which), I read Rapture studies, sports anthologies, and conversion narratives: rock and roll to Jesus, sex to Jesus, drugs to Jesus, thug life to Jesus, dulled and superficial materialism to Jesus, occasionally even the Communist Party to Jesus. One work of popular eschatology featured a low-res photo of the earth exploding; although I forget the title, the cover design is now part of my DNA. The author, whose name I have also forgotten, stands alone in a field, a bitter weed of a man, pondering, I suppose, the shocking parallels between the Antichrist and Henry Kissinger.

My real education came through the Bible—the Thompson Chain Reference Bible, King James Version, to be exact. I was given that Bible—imitation black leather, gold-edged, with two red ribbons to mark the pages—and it's still in my library at home. The Thompson Chain Reference gets its name from its dozens of appendices and glosses; to the Epistle to the Romans, for example, was appended an index containing such terms as *spiritual darkness, revelry, drunkenness, lasciviousness, wantonness, flesh,* and *lust.* From there you were directed to chain references (numbers 3193–221) in the study guide for the appropriate remedy, for example, "Self-Abasement," "The Folly of Self-Exaltation," "Subduing the Flesh," "Restraining the Appetites," "The Duty of Self-Denial," and "The Renunciation of All Things for Christ." There is a poetry to that list.

And if I didn't glean Darwinism from its pages, I did in fact learn something more important—finally, it was that Bible that taught me how to read. My Quiet Times—daily, from an early age—taught me to dig into the text, to dwell on the word or phrase, to expect an encounter with speech and its evocations. Close reading of a text, then, but also how to read the cosmos sacramentally and neighbors with compassion. All the appendices in the world can't thwart the Bible's desire to show you all of reality as a sacred landscape. When you read the Bible, the natural world becomes a place of signs, and you, the reader of the Book of God and the Book of Nature, are left helpless to do other than affirm God's beautiful creation and God's ability to speak to us through winds and rainbows and storms.

I couldn't have explained it then—the way my childhood Bible created an excess of meaning exactly because I could not imagine anything that could not be folded into divine creational interrelatedness. That excess can be exhausting, of course, as can the Bible's insistence on reading your fellow human beings as characters in the drama of a saving God. Friends, teachers, even kids who were mean to me—I could register anger at them as well as anyone, but the biblical imagination presses you to affirm that they are, as I am, sinners in need of forgiveness. What other book could you open with the certainty that God would speak to you—to the precise moment of your day, your life?

So a comfort and an intimacy with God came through the scriptures, but just like the comfort and intimacy a child might have with his parents, that comfort came wrapped around anxiety. I understood early that living a life pleasing to Christ involved self-denial and that worldly temptations lurked about, threatening my right standing before God and, indeed, my eternal salvation. When such heaviness becomes your hermeneutic, it's not long before any devotional ardor is crowded out by anticipatory dread whenever you open the Bible, because its pages can only remind you how far you have fallen short, how infinite is your failure to properly receive the superabundance of God's grace and mercy and love and judgment.

So I did learn to read in Mississippi. I learned to read the Bible as if my life hung on every word.

———

School itself changed dramatically when the first Black children arrived at Jones Junior High School in the fall of 1970. My experience of integration was of being freed from the ordeal of segregation, from a tyrannical, imprisoning worldview to one of sounds and energies. I was naive enough to think, even into my early adult years, that this was a view shared by my white classmates. But then I discovered that many of them became part of the new Republican establishment, seduced by Reagan's promise at Neshoba County. They blamed the civil rights movement and integration for the South's losing its rhythms, its religion, and its values. And yes, of course, the first years of integration brought enormous changes to the southern way of life. But I found the move to a more open society positively thrilling. Until the public schools combined, I'd never encountered a kid my age who showed any interest in ideas.

In the classrooms at Jones, a Black boy named Terrance Reed—the grandson of the prominent local minister whose home was bombed by the Klan the fall I moved to Laurel—went full Left Bank daily over literature and art, swanning through our dismal hallways with his face buried in a book.

He read everything. Terrance had no equal, and he rarely spoke an unkind word to the unread white masses he was forced to endure. Perturbed, he might shake his head in disbelief when one of us tripped over a passage in O. Henry; otherwise, he occupied an exalted perch like one of Wim Wenders's angels reciting Rilke and pitying humanity.

That Terrance had no intellectual peer disconcerted many of my friends and fascinated me—we were supposed to be the superior race, after all. Terrance Reed and his antagonist Zeke Mays were my earliest intimations of the intellectual life. Zeke, with an angular head and jaw cut sharp as a Cubist mask, banged the Black Power drum so hard his eyeballs swelled to red poker chips. Perpetually pissed off at everything, Zeke yo-mama-ed Terrance to tears with jabs about his white boots and leather grip, calling him "Uncle Tom," a phrase I knew not of, having never been assigned the great abolitionist novel. Zeke and Terrance constantly debated—NAACP or SNCC? Violence or nonviolence? Interracialism or separatism? Each morning, our cinderblock classroom in the first integrated school in the state of Mississippi turned into a salon, and I had a front-row seat, watching Stokely and Martin spar (without the affection that undergirded the two older men's arguments).

———

After a track meet on a glorious afternoon in the early spring of 1973, my father waved to me from the sidelines and asked for a quick talk. He told me that a church back in Alabama had

---

I wrote about Terrance briefly in a memoir about church desegregation; it was not until a year after the book was published and I'd moved to Charlottesville that I learned Terrance had come to UVA as an Echols Scholar straight out of Watkins High School and died a few years later in San Francisco.

made him an offer and that we would drive over that weekend to check things out. It was a church with great potential in a booming town called Dothan, twice the size of Laurel, he explained, and home to a military base, bustling regional airport, and nuclear plant. He said he and Mom had prayed and fasted to discern God's will, which signaled to me that the deal had been sealed, since God's will for his Baptist shepherds always tugged in the direction of a larger church and better salary.

Fifteen years old and coming into my athletic body, with all the weight lifting and raw egg shakes combining with a winter growth spurt to produce a lithe 5'11" sprinter and high jumper, I told my parents that I would not be moving with them. And indeed I tried to avoid the move by staying at friends' houses for the summer and tapping the generosity of their mothers and fathers. But I eventually exhausted their kindness, and my parents stopped sending me $20 bills, and thus it was that one sad August afternoon I gave up the alluvial textures of Laurel for the wiregrass region of southeast Alabama to be the new kid in town.

"Let us now go to Dothan," read the banner of the local newspaper, after the passage in Genesis 37, where Joseph—he of the technicolor raincoat—was thrown by his jealous brothers into a sewage tank and sold into slavery. Dothan was the place where I discovered that there are people who live to go bass fishing. Where native son Bobby Goldsboro returned every year for a July Fourth concert at the Houston County Farm Center. That were people who really really loved his hit song "Watching Scottie Grow." Where the mosquitos grow so big they can stand flat-footed and rape a turkey—as one deacon joked on the morning my family was introduced to the congregation. On the map Dothan looked like a mole you'd want to have checked. So I left the alluvial textures of Laurel to enter a mutating landmass with scalloped borders. In the flesh, Dothan resembled a shell-shocked vet squatting on a treeless plain drunk on pesticides. It was there that I brought to completion the glories of high school.

I graduated knowing nothing about physics, chemistry, government, or economics. My knowledge of ancient history pretty much began and ended with the indices of my reference Bible, with its chronology of salvation and its archaeological maps (marking the creation of the world in 4004 BC). But I could find my way around differential calculus, and I could write a decent paragraph; a good English teacher can take you a long way.

Mine was a recent Duke grad who had, out of some youthful idealism, returned to his hometown to road-test his very own pedagogy of the oppressed. This meant enlisting the tenth-grade remedials in a bold experiment; he would lead us through the entire catalog of English grammar as if we'd never heard of a comma (which in some cases was not a hypothetical). If this sounds like tedium, it was the opposite. It was that teacher who taught me to create on the page what my ear knew from scripture and my father's preaching—sentences with both order and dance; sentences with shape; sentences whose very structure, long before the writer made the choices of diction, could convey timbre and tone. This teacher, whom I never once saw on a Sunday morning in church, was giving us another way to sing—a set of practices that might parallel, or even beautify, the practices of the church; he was giving us another set of signs and symbols with which to make order and give praise.

Fortunately there were places in those days, and maybe some are still around, where the hermits and stargazers of the coastal South could find company. While visiting my grandmother in Jackson over spring break, I found a bookshop no bigger than a summer pantry, where I discovered the novels of Robert Stone, Don DeLillo, and Jim Harrison, which I arranged in my library alongside Graham Greene, Walker Percy, Flannery O'Connor, and Richard Wright.

And in high school, I found friends among the other guys who strived for a literary life, though we were only just learning the pretense of calling it that. On Saturdays before the summer crowds appeared, we'd fill an ice chest with fried chicken and root beer and our back pockets with paperbacks. Andy brought Camus's *A Happy Death*, Fred brought Hemingway's *Islands in the Stream*, and I had a book of interviews with Ingmar Bergman. We reclined on the sand, our bodies glistening with Hawaiian Tropic, lost in the early days of summer.

It required a steady and sometimes strenuous effort to knit the two together—my daily walk with Jesus and the restless energies coalescing around intellectual pursuits. But if I remained yoked to my Christian friends and devoted to reading the Word alongside my newfound novels and existentialist odes, it would surely all hold together. Perhaps. Graham Greene, Harry Crews, and Etheridge Knight—I wanted what these writers had. I wanted the strength to rage and howl and break on through, but I feared it even more.

———

I look back on my teenage self and want to send him off to Bennington or Berkeley. But if you were educated in states whose public school systems ranked forty-ninth or fiftieth in the nation, you would not have expected to get any advice about where to go to college. Indeed, I didn't—not from a guidance counselor or a principal, at any rate. I landed first at a Baptist college in Birmingham, and then, at the suggestion of Francis Schaeffer, who sat down with me for conversation when he passed through town on his *Whatever Happened to the Human Race?* tour, I transferred to a small Christian college in Massachusetts. I'd known the Gulf, but not the ocean until I found myself in New England, smitten by the bracing winds and rocky shoreline of Boston's North Shore.

My beginnings at Gordon College were inauspicious. Sometime during the first week of class, I was driving my lime-green Impala down Grapevine Road near the college gates when I slowed down for the tall attractive English major I'd met in Modern British Fiction. Next to her stood a missionary kid from Japan, who like many MKs hit the ground stateside like an Amish boy on Rumspringa. I wanted to make a good first impression on Sarah, and it vexed me to see the two entangled in deep conversation. Rolling down the passenger window, I started to speak when Sarah's face went gray with horror at the sight of the mangled rabbit I'd just run over, thrashing and dying on the pavement.

That leporine beginning notwithstanding, Gordon College was a revelation. In New England, the cool air caught me by surprise, as I sat in a folding chair behind the dorms, with D. H. Lawrence and Thomas Hardy in the late afternoon sun. It was the sun's swiftly falling beyond the elm trees on the far side of Chabacco Lane that stunned. The sun's warmth vanished, and, dear Lord Jesus, it was only the second week of September. In Alabama, we'd be running around the track in shorts dripping wet and sleeping with a rotating fan propped on a chair inches from our heads.

I grew to love the North Shore. I loved the expectation of studying philosophy and literature; I loved discovering Updike on his home turf. The nights were cold and brilliant and black. I made friends with a photographer who had grown up in a missionary family of Wycliffe translators in Chile, a retiring Christian socialist from Charleston who later became the editor of a neocon Catholic monthly, and a Latino man from Brooklyn, son of a Pentecostal preacher and popular radio evangelist. I gravitated to a cadre of fellow evangelical dissidents who shared a fondness for German philosophy and late walks on the lorn shores of Singing Beach. Some nights I walked into the large meadow across the street from the college and stared into the sky, dark over the twinkling village, dark over the ocean.

On Sunday evenings my father called me to ask about the weekend. He asked about my studies and the weather, but mostly he wanted to know where I'd attended church that morning. You can only lie about going to church so many times.

———

After graduating, I rented an upstairs room from a working-class family in nearby Beverly, not to be confused with adjacent Beverly Farms, under orders of the *North Shore Blue Book and*

Social Register. Beverly Farms was a playground to the Myopia Hunt Club set, palatial homes hugging the rugged coastline and their starchy inhabitants. In Beverly, the O'Hanlons welcomed me into their cozy, wind-battered bungalow as if I were kin, inviting me down for meals or to watch the Pats on Sunday afternoon, sometimes to say the rosary with Mrs. O'Hanlon.

I was only there for a few months. Come fall, I'd enroll in divinity school, but for now I was taking a poetry class through Harvard Extension and drinking each night with a classmate who'd go on to publish in the *New Yorker* sixteen lines that locate Jesus and the hemorrhaging woman in a New England grocery store. It's a poem of such power that I'll be reading it, I'm sure, till I die. But Andrea must have thought I was nuts, night after night in her own rented rooms, enough beer and bourbon between the two of us to float a barge. My desire for her, for the life those poetry classes seemed to promise, only made me drink more. I couldn't consummate either one— neither the literary life nor sex with Andrea.

I was trying to make a go of writing (I'm sure I thought the hangovers were ingredient), sitting at my desk each morning for three or four hours, scribing verse, and sometimes returning to the novella I'd written as a senior thesis—a picaresque devoted to the hobo H. Rufus as he traverses the Piedmont Plateau into the parched woodlands and the coastal plains of southeast Alabama, where, smelling of sweat, tobacco, pine sap, and mystical heat, and hounded by fears and phobias of the kind likely to arise during periods of complete loneliness, he is seized by grace and made whole.

I worked in the afternoons and evenings with a youth group

of Black and white boys and girls from Lynn ("Lynn, Lynn, the city of sin," the conductor of the commuter train always said when we pulled in). I loved those kids; they were sweet and wild. This life seemed then, as it even seems now, to hold together all the goods I cherished—poetry and thought and Christian faith that ought, I was beginning to see more clearly, find expression in relationships of trust and acceptance, with people who heard the gospel as a still point in their chaotic world and who then returned to their families and neighbors with a sense that the Lord might be there too, as a brother, sister, and friend, "turning shadow into transient beauty." I don't know what became of these spirited children of inner-city Lynn. By the end of the fall, the church's pastor was caught in an affair with a parishioner, the pastor's wife filed for divorce, the parishioner's husband came after the pastor with a tire tool, and things kind of fell apart at the Tired Pilgrim Baptist Church.

I didn't see any tension between the mornings at the desk and the evenings with the youth group (and the tension I saw between those goods and the drinking I blotted with more drink). I brought my guitar to the Baptist church basement, and the kids and I sang James Taylor and Cat Stevens, we studied the scriptures, and I tried to engage them with Christ through my love of literature. One evening, I led them in an exercise around T. S. Eliot's "The Waste Land," interleaving it with a meditation on Colossians—what both texts say about suffering and about the times of transcendence possible in the midst of that suffering— and halfway through I understood I was being watched.

I mean that literally: the church's pastoral intern—a seminary student, maybe two years older than I was, but endowed with all the authority of one who (let's put too fine a point on it, shall we?) was becoming my father—had crept into our basement and was observing the lesson, observing our prayer. After the kids trickled out, back to their gray-shingled row houses and stoops, young pastor Mike shared with me in no uncertain terms that he was deeply concerned about my hermeneutic. You can imagine how this went, this confrontation between two pious, eager, cocky, insecure young Christian men, one about to don the mantle of ordination, the other who had read his Spinoza and was utterly certain that his opponent had majored in electrical engineering or accounting at Wayne State and didn't know what *hermeneutic* meant.

My hermeneutic, Mike said, didn't seem to be grounded in the Word (by which he meant the Bible—circumscribed by the aggressive dictates of biblical inerrancy and the new brigades in the battle for the Bible). Calmly, I explained that to correct my worrisome hermeneutic would require him to ask questions about epistemology, theories of validity and meaning, and methods of textual interpretation. "Neither your discomfort at hearing the lost teens of Lynn singing along with my Cat Stevens covers nor your unease at hearing me recite poetry alongside the epistles of Paul constitutes a critique," I said.

I seemed fearless in my response to him, but I *felt* devastated. What Mike had offered was a formally perfect pastoral rebuke. And if you are among those who care enough about the life of faith to be in a position to receive such castigation, the pastoral rebuke is experienced as a laceration of your most intimate offering, your most intimate gift.

Pastoral rebukes were coming in long letters from my

father too, telling me how much God had planned for me and at the same time how fragile the sense of call—how fragile the blessing: "You will be constantly under assault by Satan," he wrote. "You must maintain constant vigilance." As the fall wore on, that epistolary intervention turned into equally long, troubled phone calls. My parents were "worried about my priorities." When they learned that an old girlfriend from South Carolina was coming to see me for a weekend, they went nuts.

The morning hours at my desk were meant to be the point of all this, but I began to conclude that the people I was reading—Virginia Woolf, T. S. Eliot, Camus, Dylan Thomas—had in their psychic makeup something that I simply lacked—the capacity to give myself fully and freely to a creative venture. Specifically, to the creative work of poetry and fiction, to such writing—as I experienced it, alone in my room—that confronted me directly with libidinous desire, without the protection of the third person or the filter of scholarly distance. I had expected the bodily experience of sitting at a table writing—blocks of time now open, without the demands of classes and papers—to be thrilling, and of course it was. The pose of receptivity was like a dangerous unleashing, which meant physical arousal—first the erotic, coursing through pencil and paper like some kind of club drug, and then the inevitable tightening of breath. Which is to say it was dangerous, and I was not certain I was equal to the danger.

One afternoon, sitting in that rented room of that working-class Irish Catholic couple on the North Shore, a daydream. It was a daydream of living in Paris or New Orleans, immersed in the pleasures of my bodily life, and that, of course, was the rub. Meursault had his Marie—had sex with her, I mean. Binx, his

Kate (his Sharon, his Maria, his Linda). I had dalliances with women that may have led to orgasm but could never lead to penetration and, in some essential way, never really led to touch.

Does this seem absurd, from our vantage point four decades later, this picture of a man so bound to his connate vision of the good life that he couldn't get laid? Or does the strangeness lie not in the virginity, but in that young man's refusal—inability—to disappoint his parents? Perhaps I am writing this book to show, shorn of the absurdity, the elemental choreography of it all.

In my childhood, they used to speak of smelling depravity. You could smell if someone had had sex, and they didn't mean the smart tang of womanhood; they meant the darkness of transgression. What I knew, those nights with the poet and the booze, is that if I had sex, my mother would know, and she would somehow be undone, this woman whose mental equilibrium seemed to demand the purity of her only child. What I knew is that such transgression would remove me from relationships I felt I could not live apart from, I, who needed the ancient drama of calling out for my mother at night, the drama of her coming to me and lying down next to me. I was oppressed by their protection, and I hated it, but their presence was also a necessity.

I fantasized their death, or my own. I felt fairly certain that if I transgressed—if I let down my guard and fornicated with Marcia, Linda, or Sharon—I'd have to kill myself. I felt fairly certain that the only way to find creative freedom, the breathing room to write, was through my parents' death. I write now, how many years shy of their deaths? Two? Three? Ten, if we're all lucky. And having survived the virgin suicide that for so long felt my own proper fate.

PART III

Loyalty to the Event

HDS, Redux

Harvard Square, in the promise of fall. I fairly staggered under the swoon of possession. Whatever doubts I secretly harbored about my capabilities to flourish—such doubts seemed to me, as the fall semester of 1981 commenced, but garden-variety jitters amid the excitement of summer rushing toward a new school year.

And then the breakdown.

A few weeks after the cataclysm in my dorm room, I ran Quentin Compson's path to the river. Quentin, the Harvard student who jumped to his death in the unforgiving waters of the Charles River in William Faulkner's *The Sound and the Fury*. Quentin, the original southern sad boy, who traveled north to the Ivy League on the profit of his father's farm and who, once there, tied flatirons around his ankles as mayflies skimmed in and out of the shadows above the surface of the water. "The reducto absurdum of all human experience." Are other fictional suicides memorialized with plaques? "Quentin Compson III. June 2, 1910. Drowned in the fading of honeysuckle." During my time at Harvard, the plaque was a melodramatic destination site not only for wayward Southerners.

Why had Quentin gone to Harvard? Why had I—trapped in the narrow corridor of Rockefeller Hall or what it now signified, a gray half-lit future where all things that had once been stable enough were now grim, twilight, chimerical—not drowned myself as well? What was it that led me stumbling down Dunster Street?

As it happened, the weekend before the breakdown—that weekend with K at Elisabeth Elliot's house—I'd gone for a long run, as I often did when I needed a break from studying. On this day, I'd followed Maple Street to the Pingree Woodlands and a trail across a vast expanse of wetlands, a trail that only a month earlier I'd run beneath a high coastal sun sheltered by white pines and beech. But now the woods were shattered and fingery—there were no leaves, and there was a sense of freefall to the run—it got cold and dark—halfway into the run I realized it was overcast, and then gray sheets of cold rain came down on me. Later, I would think implausibly that this was foreboding, that it was premonition or a spell cast. Though when K asked me how the run was, I said it was fine, and I made a joke about the rain. Later, I would think that I was on a journey, and I had expected it to be filled with light and with the ever expansive effects of breakthrough, but in fact the journey took me into the groves of darkening trees.

K was the daughter of a Presbyterian pastor of an affluent church on Peachtree Road in Atlanta. We'd met through friends of my parents—friends who knew that a beautiful woman who'd turned down Duke for Wheaton would pique my interest. K very definitely had, with her wit and her faith and her beautiful breasts. Though at first we simply wrote amorous letters across time zones while drinking coffee and smoking cigarettes and making eyes at and making out with people nearer by. But often I found myself thinking about her while chatting up a woman I'd met in class or a party, and I found myself thinking, for the first time, about marriage. Yet always the ridiculous hunger to read everything, to write beautiful sentences, to stroll through marvelous galleries in leisure hours, to be seduced by and to seduce, and in turn to taste and eat and fuck, the Iranian postmodernist, the medieval historian, and/or the mogul's daughter in the deep dark love dungeon of Widener Library.

And now, weeks after the breakdown, I was running the path by the Charles River. I was in this body that I had once trusted, the body that had not many years before been trusted and strong. I narrated myself as I ran: *I am taking this run through the streets of Cambridge, and it is neither restorative nor self-forgetful in the roar of urban traffic.* By the time I reached the boathouse near Anderson Bridge, I felt the first clutch of an asthmatic wheeze, of *asthmathe*, "panting breath of air to seek," as I'd learned in Greek that fall. This would be my first asthma attack since junior high. I thought I'd outgrown it. But here I was again, panting breath of air to seek.

————

Let's step back for a moment from the gaping sorrow in the House of Compson and take stock of the public forces weighing down on my situation, because to some readers my anxieties might seem an appropriate response to the events of the time.

Ronald Reagan took office in January of 1981 with the promise to make America great again. Watching the National Republican Convention the summer before with my Republican cousins in Mississippi felt like being in a room with baboons. Every jingoistic platitude was met with hoots and squeets. An unbelievable fortune was in the offing, some repristination of our lost valor. Vote for the Gipper, and the South would rise again, and its name would be America. From the bayous of Acadiana to the Tidewater lowlands came a resounding "Shazaam!" Soon the Dutch and his Pretty Nancy were bedding in the White House, and the nation's skies cleared to tax cuts and binge shopping and lavish indulgence. I'd never much discussed politics with my friends back home. If I had to reconstruct our Christian social ethic (a term we never used) or worldview (the term we preferred), I would say it comprised some strange gumbo of Jesus-freak pathos, Oswald Chambers austerity, and Freebird libertarianism.

From the banks of the Charles River, I stood appalled at the national moment. The old friends who made it big selling insurance to truck drivers or repurposing strip malls. United by quick fortunes in the strangest of ventures, some flew to London for the last of the Cabbage Patch dolls; others converted beach bungalows into steroidal buildovers on quarter-acre lots. Who knew you could make a fortune selling trophies and corporate knick-knacks? Apparently Marvin Aberfoil, an amiable pothead with a heart for Christ, who bought a house on Lake Eufaula before his thirtieth birthday. Viva MA Quick Trophies, Awards, and Plaques—just off Highway 431 at the Headland Exit, Toll Free Number. 800-FUCK ME.

It was morning in America. Bye-bye, gridlock; make haste, malaise. It doesn't get any better than this. Nights belonged to *Dynasty* and *Dallas*, cocaine and clogging (what the hell?!), to trivial pursuits raised to a civic sacrament. The "era of good feeling" passed me by like some peripatetic god, like the sad scent of honeysuckle in late summer. I chose dissipation instead. A good thing if you have to choose between the two. I chose haunted male characters who renounced the world and paid the price.

———

In winter, I visited my parents, who had moved to a brick Cape Cod on a cul-de-sac in northwest Atlanta. But even there—home (their home, if no longer exactly mine)—I was beset. In the four-poster bed I'd slept in since childhood, pillows coiled around my head, I struggled to mute the sound of barking dogs. Their back yard descended into a gulley where the city limits ended and the county—Cobb County—began. The county was named after the despicable senator, slave owner, and secessionist Thomas Willis Cobb, not, in fact, after the rapist Popeye and his corncob dildo in Faulkner's *Sanctuary*, but if the county calls to mind Popeye and/or Temple Drake (she, the victim of the rape), you'll find it easier to understand the terror caused by the neighbor's fenced-in compound of feral insomniac hounds. I wanted to shoot them all dead.

The barking grew to terrific howls as I lay awake. Until I resorted to my own primal needs and called out for my parents. It was my father who obliged that night. I'd turned on the bedside light to give the appearance that I was up late reading. When he arrived at the door, he grabbed a college football magazine from my desk and sank into the bed beside me, and we were both soon out cold.

Yet the formula that had always steadied me in the past—jetting home to Georgia for my mother's gumbo and cobbler, watching a football game with dad, binging novels and trash TV—now failed to deliver. I didn't know a soul in Atlanta other than the good souls of my parents. I flew back to Boston feeling stupid and ashamed.

Not long after I returned to Cambridge, a book arrived in the mail, looking road-worn in a cracked Mylar cover, *Run and Do Not Be Weary*. Two twenty-dollar bills were paper-clipped to a note that read, "Just a lagniappe for a meal or outfit." Bless their hearts.

I didn't read *Run and Do Not Be Weary*, though the book has somehow managed to appear among my belongings in every town I've lived in since the end of the Carter administration, most recently in a lawyer's box in our basement. Three inches of water left by an apocalyptic rainstorm forced me into a morning of hard labor, setting up a sump pump and hauling crap out to the curb.

———

By the end of the spring semester, it had become impossible to imagine that the symptoms would go away. I couldn't picture the conditions of a cure, how it would look or where it might come from. "Dig deeper, young man," I could hear the saints drawl, their words falling like sorrows from an amber cloud.

I bore down on my calling and prayed for the mercy to endure. I prayed even more to forget. For the years when I journeyed more or less unfazed through the particularities of me. Or whatever it was I'd had before the breakdown, the mercy of waking stupefaction. I wanted it all back. The grace of unknowing. Animal life. I wanted what the thing Cathy Park Hong fantasized about in her memoir *Minor Feelings*: "someone to unscrew my head and screw on a less neurotic head."

Every defense failed. My symptoms felt concentrated into permanence. I had lost the capacity for happiness. I was the bitch of random freak-outs.

My cartographies of avoidance grew ever weirder. Libraries and lecture halls earned gold stars; runners-up included low-church sanctuaries draped in melancholic reds, newly cleaned houses carrying sad scents, and (as I have tried to explain) subterranean cafés and poetry bookshops.

Also the theater. I had snagged tickets for *Waiting for Godot* at the American Rep—Andrei Belgrader's production, which would bring Beckett's affections for vaudeville front and center. The cast looked amazing: Tony Shalhoub and Richard Spore as Pozzo and Lucky; John Bottom and Mark Linn-Baker as Estragon and Didi. *Godot* did not need to be a total downer; it could be cheeky and biting comedy, and that sounded very good to me. I'd bought front-row seats shortly before the deluge.

By opening night, I'd become a man for whom pleasure opened only to terror. It came as no surprise that when the house lights dimmed, a tremendous panic leapt out of the desert darkness. The thought formed sadly against the pantomimes of death, *I can do all things through Christ who strengthens me.* Sighing and swaying, I monitored every heartbeat and breath. I wanted to howl. I wanted to be somebody else. I wanted to run into the cold, abysmal night and disappear.

Meanwhile, Shalhoub and Linn-Baker brought down the house with their magnificent insults:

> *Ceremonious ape!*
> *Punctilious pig!*
> *Moron!*
> *Moron! [again]*
> *Vermin!*
> *Abortion!*
> *Morpion!*
> *Sewer-rat!*
> *Curate!*
> *Cretin!*
> *Critic!*

I hardly noticed.

———

"When one is young," Nietzsche writes in *Beyond Good and Evil,* "one venerates and despises [the world, life] without any art of nuance.... Later, when the young soul, tortured by all kinds of disappointments, finally turns suspiciously against itself, still hot and wild, even in its suspicion and pangs of conscience, how it tears itself to pieces impatiently!" Punishment takes the form of mistrust against instinct and feeling; you torture your most intimate enthusiasms with shame. And above all, you create principles and systems and use them against youth as a punishing rod.

And so it went year after year, the terrors unleashed on a night in Cambridge turning into my very own *Krankheitsphase.*How nearly impossible it is for healthy people to understand the sadness at the core of the anxious person. Whatever mental tricks remain useful, whatever religion offers pep to the normal and the adjusted, is temptation to the one who fears life and for whom "every sensation is oppressive."

Belief was not my problem. Belief was a habit long ingrained in me. You can believe on the strength of the absurd. You can wager on God. Number me among the freaks in the plum trees, if you want; I don't mind. The stoners called me a Jesus freak in high school, and I liked it. "God so loved the world." "It is for freedom that Christ has set us free." "He leads me beside still waters." I've tried, and I've failed, to speak another language. Although every word, any word, evokes some irresistible luminosity.

But this flesh? Redeemed or not. Again?

. I borrow this term from Erik Erickson's discussion of Martin Luther's mental health: "the years in which we are most interested—when Luther was twenty-two to thirty," writes Erickson, were "part of one long *Krankheitsphase*, one drawn-out state of nervous disease, which extended to the thirty-sixth year; these years were followed by a period of 'manic' productivity, and then by a severe breakdown in the forties." *One long . . . drawn-out state of nervous disease . . .*

Christian Anxiety:
A Short Theology

When the protocols of biblical self-help fall short in the treatment of mental illness, as they inevitably do, anxiety and depression will hunt down vulnerable regions of the psyche like an angry infection attacks nerve and muscle. You've seen the reports. How men turn to drugs, porn, and power, to cruelty and messianic ambition, and religious abuse masquerades as God talk. Meanwhile, even as the number of Christian suicides increases each year, nearly a third of all evangelicals believe that those who take their lives will go to hell.

This is the plight of the evangelical self unaccompanied by respect for the mind's intricate dramas: the salvation event shatters. The evangelical self requires its own abasement. I speak of my body as the synecdoche of sin.

In his psychiatry textbook, Joseph Lévy-Valensi described anxiety (*anxiété*) as a dark and distressing feeling of expectation. *Anxiété* comprehended the psychological and cognitive aspects of worrying. In contrast, anguish (*angoisse*) was defined as the experience of spastic constriction of voluntary or involuntary muscle fibers—as when my legs refused to walk normal or my tongue became an entity with a will of its own threatening insurrection. *Angoisse* could be experienced as bronchial spasm, shortness of breath, intestinal cramps, vaginismus, urinary urgency, pseudo-angina pectoris, or headache. Lévy-Valensi was stripped of his sinecure in Paris, deported to Auschwitz in 1943, and murdered.

Another way of parsing the terrain is to distinguish among three types of anxiety. The first we might call *practical anxiety*; this is anxiety that is useful, instrumental, the kind of fight-or-flight instinct written into our genetic inheritance—our evolutionary inheritance. The second is *existential anxiety*; this is the experience of dread, the abyss, the nothingness. Here anxiety assumes a conceptual form; as when one stands in awe before Picasso's *Guernica*, encountering the human condition with no cover, raw, disruptive, aimless, and bereft. The third type is *pathogenic anxiety*; this is the beast. This is anxiety that obliterates the powers of reason. Practical anxiety and existential anxiety may contribute positively to self-formation, growth, and individuation, but pathogenic anxiety takes functional capabilities away from its sufferers and overpowers them by the blunt force of the symptoms.

Theologians, it's worth noting, rarely consider anxiety in its first and third aspects. The theologians who tend to anxiety at all consider it foremost an existential category. Many of these accounts invoke Martin Heidegger and Søren Kierkegaard; anxiety proves to be a portal of experience through which the individual is awakened to the perils and risks, the fragility of human existence.

———

I wrote earlier of the episode commencing out of a high pandemonium descending with pinpoint precision. Literary accounts of madness in which madness is personified as a beast or a creature may evoke anticipatory fears similar to actual symptoms. These narratives of shock and awe—sometimes spiked with boasts and bravado, "I am so damaged, it's gonna blow your mind!"—activate an interior searchlight that scans the psyche for easy prey. Laser sharp. But that is not my story.

As I uneasily approach the neurotica section in my library—which I reorganize every few months and which now resides in four industrial bookcases in the basement—the black and red cover of Daniel Paul Schreber screams for attention the instant I flip the light switch. Schreber was a sad, deranged nineteenth-century aristocrat who was confined to an asylum and wrote an account of the visions and mysteries that were revealed to him in psychotic episodes, *Memoirs of My Nervous Illness*, a title that does not adequately prepare you for the book's unabated horrors. Nor does the original German, now that I think of it, *Denkwürtigkeiten eines Nervenkranken*. I don't know how the book ends, because I've never finished reading it; it's astonishing, but too harrowing to be borne. I own five different editions. I keep thinking that if I find an edition that feels good in the hand, I'll stay the course.

Then Jamison. If you've read Kay Redfield Jamison's memoir *An Unquiet Mind*, you are not likely to forget the details of her first major psychotic episode. One evening early in her tenure as a psychiatry professor at UCLA, Jamison stood alone in her living room watching "a blood-red sunset spreading out over the Pacific," when she suddenly felt a "strange sense of light" at the back of her eyes, and from "inside her head" she saw a "huge black centrifuge." It was at that moment, she writes, but outside her now as an independent entity, "a tall figure in a floor-length evening gown" approached the centrifuge "with a vase-sized glass tube of blood in her hand."

I read the passage again today. A centrifuge whirls and splinters into a thousand pieces; everything is covered in blood. It's the stuff of B-list movies. At least that's how I'd prefer to see it, as monsters on a screen. How difficult it is to understand another person's madness. I don't even want to try. The madness I understand is my own: how it feels to wake up on a skeletal cot in a universe of dread.

Hans Urs von Balthasar, in his book on Christians and anxiety, writes that in episodes of terror every superlative transcends itself over and over in a dialectic of ruined ends. "The smoke is an animal, the locust is a scorpion, the scorpion bears the marks of every predatory beast," he writes. I know how this feels too: thesis, synthesis, collapse, repeat. How each episode "dissolves into a surging sea of flames, each one of which nevertheless strikes home, in precisely the manner least expected." Not in the likeness of gods or animals—an animal you could tame or kill, a god you can appease—but mute constricting flesh, world without end.

This is the reason so much writing about anxiety appears to the nonanxious like a compendium of medical facts. The anxiety event resists the anxiety idea as if it were as ineffable as Yahweh—unless you're inclined, as many anxious people are, as I am, to surveil every incoming word as a potential assault on the mind and body.

Reading doesn't always plunge you in. Sometimes I read in order to stand outside the event. On those days, I don't read Jamison; more likely Kierkegaard, in whose company one entertains the claim that to imagine anxiety as freedom's possibility is to have survived its unmaking. To think about anxiety as a force opening the soul to truth is to have made the leap beyond it. "Anxiety ennobles," Kierkegaard says. "If a human being were a beast or an angel, he could not be in anxiety. . . . The more profoundly he is in anxiety, the greater is the man."

Kierkegaard's brooding mediations are luminous but philosophic. If you read them against the tradition of transcendental subjectivity, as he wrote them, anxiety appears as the force of eccentric existence taking down the system. And a damn good one. Kierkegaard posed the anxiety idea as a counterlogic to the Hegelian dialectic, arguing that Hegel and the German idealists, in creating such exquisite philosophical systems, gloss over anxiety, wrath, suffering, and anguish with the optimism of the age, that the experience of dread and its attendant terrors is erased. For the Hegelians, torments within and without become little more than "the deity's creative birth pangs." Where is the "positing of sin"? Kierkegaard asks. An excellent question. Against the eviscerating effects of the system, you say, the anxiety idea speaks the truth of the simple, unassimilable "I."

This is the anxiety idea. The remainder is the particular

individual, sitting at his desk on an ordinary morning in Copenhagen, in Charlottesville, that enables him, me, to speak: "I am this one here."

But Søren, where in your liquid thought with its undercurrents of irony and spiraling wit may I find an anxious body?

In the anxiety event—that's where I find it. The sadness of being alone is exhausting: living at the mercy of random thoughts, being afraid to travel alone, the dread of twilight. The night that left you gasping for air, when the walls shook—the assault that came, or seemed to come, out of nowhere, the roaring darkness—the night that holds your misery, that you later tell your shrink about (hang on, he's coming) but rarely your family or friends—does not reveal its powers at once.

———

I think about that preacher who hoped that everyone in the congregation would have a complete nervous breakdown. For real religion is this: to stand ever and always at wit's end, empty and abased.

But don't mistake our self-loathing for passivity: I speak here of white Christians with material resource and bright futures. We shall march in service to whatever aural powers have been certified as the real deal. The signs are everywhere. In hypomanic fears of extinction. In the narcosis of purity. In the hysterical claim that we are always being persecuted. It does not take much to shed our skin. So tread carefully. The pitch for the God-shattered ego is finally a grasp for power and control, and it must be resisted for the sake of our created integrity.

The Reformation gospel that we are saved by faith alone was too good to be true, so we turned the vicarious substitutionary atonement into a verger's rod.

"The crucified God draws near to every person in their experience of forsakenness," the theologian Jürgen Moltmann writes. "There is then no need for the Christian to think that self-destructive self-accusations enable the justification of sinners." Otherwise, would not God's free love become dependent on our own capacities for shame?

PART IV

Testimony

Charlottesville:
The First Sojourn

K and I honeymooned in Bermuda at a cottage in Somerset shaded by palmettos and black mangroves. It was late in the summer, and the population of tree frogs had gone quiet for the season. A solitary peeper atop the air-conditioning unit in the window accompanied our timid explorations. I'd lived my life in anticipation of this week. We had explored each other's bodies before marriage with a few compensatory movements toward the spirit of the age, but we were both virgins according to the letter of the law. We had waited for the one who would make us complete.

Engraved in my wedding band is a line from Auden: "We are linked like children in a circle dancing." And in my wife's are the words: "My dear one is mine as mirrors are lonely." Our expectations far exceeded anything our untrained bodies could deliver.

You might have heard the story of how the writer John Ruskin was so horrified by the sight of his bride's pubic hair that he failed to achieve an erection on his wedding night. To my knowledge, no one has ever said that his Effie's chaparral was unusual in size or volume. It was just the sight of the bush; he wasn't expecting it, and didn't like it when he saw it. And although I would not say this was true of my own first encounters on the afternoon of Saturday, August 7, 1982, in a junior king suite at the downtown Atlanta Hilton, my knowledge of the landscape below was mostly gleaned from impressions formed in reverie.

When, during one especially turbulent season, I confided to a youth pastor the profound sexual fantasy into which I often fell—of such intensity that I could lie in bed late at night listening to the rock station in Panama City and ejaculate into my down comforter without touching myself—he hurried to explain that this was expected in a young man's struggle. Depressions would come and go until I met the girl God has been preparing for me, and if I persevered in purity, the desires of my heart would be fulfilled in the conjugal union with my betrothed. The explanation sounded reasonable enough. "Take delight in the Lord," one of the first Psalms my mother taught me, "and he will give you the desires of your heart." All the way down to the irksome particulars. I trusted the pastor as I trusted my mother; the Word never lied.

It wasn't until I married that my desires found their true métier in the singular melancholy of the virgin's wedding night. With our vows now certified by God and the state of Georgia, K and I had finally entered into the covenant that allowed us to make love without sinning, in any position, as far I knew. Someone had whisked us away from the alcohol-free reception my new in-laws had convened in the cafeteria of a Christian high school—which also included neither live music nor dancing—and deposited us outside the corporate hotel downtown. Whatever festivities had been shared over church-lady punch and vanilla cake on that August afternoon quivered like a strained smile; and soon the parking lot was empty. K and I lay naked and spent with the curtains drawn and an hour left before the hotel's restaurant opened for dinner. That's when I began crying.

I was always reassured by the promise that a girl my age (maybe younger) mysteriously traveled a parallel track, surrounded by her own cloud of witnesses, on her journey to me. Our bodies would not clash and collide. They would become a lamp that glows and warms. Of course the fantasy was more than our bodies could bear.

———

Mere months after marrying, we moved to Charlottesville. K gave up a fulfilling job at the Boston Children's Hospital, so that I could embark on a doctorate in philosophical theology. We'd be in Virginia for six years; we had no way of knowing that eleven years after leaving, we'd return, that this grad-school sojourn would be remembered as our *first* sojourn in the town where we'd wind up raising our children and easing into the rhythms of middle age, the town where, it seems likely, we'll one day be laid to rest, buried in the shade of a white oak tree in the university cemetery. All we knew when we arrived is that I would complete the education that would somehow secure our future.

Doctoral studies in the humanities requires, and presumes,

a kind of fevered reading like no other. Since the breakdown, reading had become laborious—once a pleasure, now bound to my illness in ways that confounded and vexed me. So discipline became my drug of choice. I retained the ability to warble through a written page by means of elaborate time-management protocols and assiduous notetaking. Useful, if not essential, habits of scholarly life, unless you live each waking hour in obedience to a Zassenhaus fifty-five-minute timer and a day planner that reads like Josef K's to-do list.

My professors aimed to march us through the seminal texts of nineteenth- and twentieth-century theology, disproportionately represented by Protestant liberalism and especially German philosophy. We were taught to recast the Christian faith and its truth in terms pleasing to enlightened moderns. The story of modern theology, which I was to absorb as I transcribed hours of lectures meticulously into my notebooks, understood the doctrines of the church to be meaningful as lessons about human experience, but little more. The great Berlin theologian Friedrich Schleiermacher may have grasped how this flawed tradition could sometimes produce inspired results, as in his exquisite rendering of religion as "a sense and taste of the Infinite." But it took Feuerbach, Freud, and Marx, the "prophets of extremity"—as my colleague Allan Megill calls them—to better understand the repercussions of Enlightenment skepticism and Romantic anthropomorphism. It was but a matter of time before the death-of-God theologians of the 1960s, with their pipes and paisley ascots, cast the old metaphysics aside like last week's church bulletin and, in keeping with the zeitgeist, embraced the void, the nothing, the abyss, the infinite jest all the way down.

I read all this arid Protestantism even though my intellectual interests steered me in the direction of modern Catholic writers. Most Southern Baptists grew up thinking of Catholicism as a deficient form of Christianity, but that was not really my story. I thought of Catholicism—when I thought about it at all—as something like an eccentric uncle. I'd grown up with its allure strangely tolerated by my parents.

Our home in Mississippi was an easy 150-mile drive down I-59 to New Orleans, and one of my parents' whimsical indulgences was to honor my occasional plea to skip school and go to the Crescent City on an unplanned weekday jaunt. How my parents occupied themselves on those urban afternoons I don't know, but they let me, at age thirteen, wander alone through the shops and cafés of the French Quarter. I loved the colors and the textures, the dank air and the bewildering smells—the strange beckonings of jasmine, alcohol, beignets, étouffée—and not a whiff of Baptist piety. I loved the way my body seemed almost to float in a sweet green heaviness, so different from the suffocating effects of Laurel.

When, at regional basketball tournaments or track meets, I'd chance upon girls from New Orleans—sassy, beguiling Monica Vittis in cutoff jeans and flip-flops, voluptuous daughters of Ursuline, Sacred Heart, and (heavens to Betsy!) Our Lady of Divine Providence—I allowed that they had found God too in their own way, but that, in and through me, they would be led closer to Jesus than they ever thought possible—

seriously. I could do this for them, I felt with an aching ardor, with a zeal equaled only by my desire to get in their pants. I'd knot up at the thought of returning with them, somehow to disappear into their lives.

There was never any question of my throwing my own ecclesial oar in with Rome (cf. "deficient form of Christianity," above), but it's no surprise that my aesthetic loyalty to New Orleans, my spiritual and erotic loyalty to New Orleans, led me to read Walker Percy in high school, which in turn led me to Graham Greene and Flannery O'Connor in college, all of which had me thinking, my first year in Charlottesville, that perhaps the new subdiscipline "Religion and Literature" might be where I'd make my scholarly mark.

Or perhaps it would be among the medieval mystics, whom I first studied in a seminar under the tutelage of the brilliant Carlos Eire. Julian of Norwich, Meister Eckhart, Bonaventure. They were, of course, a revelation—how could they not be after years of Oswald Chambers? Oddly, the mystics seemed to prefigure much in the tradition of evangelical self-emptying— their stress on voiding the self before God and for God felt only familiar. But the mystics came to, and sometimes traversed, a threshold. They described entering into divine life in a way the pious evangelicals of postwar America never had. I read them, and I wanted that kind of breakthrough—that apotheosis.

But neither *The Cloud of Unknowing*'s elusive monastic guidance nor Meister Eckhart's apophatic transcendence ultimately proved an intellectual home. Rather, it was in the lush, free spaces of the Berlin Grunewald that I found, midway through graduate school, the thinker who would become my closest intellectual companion: Dietrich Bonhoeffer.

K and I lived in a two-bedroom townhouse in an apartment complex on the outskirts of Charlottesville, and I claimed the second as my office. There, in a Windsor chair my parents had given me as a graduation present, with my notes and papers arranged in neat piles on the pub table I'd made my desk and a framed print of *Starry Night* just at the edge of my peripheral vision, I arrived finally at my comprehensive exams. It would not be too much to say that in the midst of that trial, a path cleaved open, because it was for those exams that I first read Bonhoeffer.

Bonhoeffer's participation in a plot to kill Hitler and his subsequent execution by the Germans—heroic though the story was—wasn't what most affectingly spoke to me on those Virginia afternoons. What I began to see in Bonhoeffer is that the way to Jesus is through descent into the complexity and the brokenness and the joy of the immanent world. There is no grace without obedience; no freedom without the weight of the cross; no "Neither do I condemn thee," Jesus's words to the woman taken in adultery, without "Go and sin so more." The terms of grace are quite explicit. I began to see in Bonhoeffer that the way of Jesus was at once the way of suffering and desire.

There in the nondescript townhouse, with my wife a few steps down the hall, I read Bonhoeffer, in the letters, insisting that one dare not speak of Christ's resurrection of the dead before one has personally felt and experienced not only the sorrows but the joys of worldly life. *How was Bonhoeffer like and unlike the mystics?* I remember thinking as I prepped for exams. *The end was the same; the end was joy; the end was incomprehensible beauty and reconciliation. But there was also pleasure and heat and the bodily senses, and in Bonhoeffer you found God not by ascending, but by plunging into the depths and desire.*

On one of his first journeys to Spain, he noticed how the air held the fragrance of almond trees and lemon and the sweet saltiness of the sea close by. I'd been immersed in northern European abstractions, wholly shaped by concepts without flesh, let alone olfaction. How revelatory to read a theologian moved to address "the fragrance of your being," to address the botanical, to address place.

That first afternoon reading Bonhoeffer, I didn't know—how could I?—that I would live for decades *auf den Spuren von Bonhoeffer*—on his trail, following his traces. Yet, even then, Bonhoeffer began to help me understand that, in an animated, lively sense, Christ is the mystery of the world, Christ is at the center and in the margins, Christ is in grandeur and anguish and pleasure and bliss and creative freedom, and that because this Christ is the source of life, we are relieved of the need to concoct some existential or conceptual master plan for our own salvation. That idea can be abused, of course. For Bonhoeffer, it is disciplined by being framed within a tradition of Christian belief, but in his thinking an aspect of autonomy is nonetheless accorded the individual. There were those in Bonhoeffer's Lutheran tradition who thought that being saved eviscerated the saved one's autonomy, but Bonhoeffer judged that nonsense.

He believed that affirming Christ as Lord enabled people to assume their specific and idiosyncratic aspirations and desires.

In time, it was Bonhoeffer who helped me to see that life with Jesus leads disciples toward the world; that "it is only when one loves life and the earth so much that without them everything seems to be over that one may believe in the resurrection and a new world"; that confidence in the gospel can and should permeate the self so that Christians might relate to the world out of thoughtfulness, courage, and humility—not out of dread.

But in the claustrophobic townhouse, those were just glimmers, and dread was still very much the order of the day.

———

Our fourth year in Charlottesville, I was taken by an attack that rivaled the breakdown in Cambridge, except perhaps it was worse, because the symptoms—a perpetual state of panic, everything fraught with panic and fear—were unabated over days and weeks, refusing to respond to the tricks and feints I'd learned since Harvard. I tried herbal remedies and cognitive psych tricks—*Rethink expectations! Terrified of attending that conference in Amsterdam, where you might be devoured by lust and desire? Just remind yourself that you don't have to go to Amsterdam! Put your terrors in an imaginary cabinet and know that you can visit them each day if it would soothe you to do so.* I drank myself to sleep each night, but knew wakefulness more than sleep—I wanted a German woman who was a friend of K's and mine, she wouldn't let me, and I didn't think I could pull it off anyway—I couldn't sit still—I couldn't work—I was frightened of everything. None of the old momentary reassurances (the belief that there were certain places or certain times of year that would offer a reprieve from the ferocity of the symptoms) reassured.

I was assigned, by a small local journal, to review a book, something of a coup for a young scholar just setting out. When I opened the envelope and saw that the book treated the Neshoba County murder of three civil rights activists, the very murder whose agents had been on trial the year my family moved to Laurel, I first thought, simply, *How odd*. I knew I'd never write the review or even read the book. I couldn't. Every time I tried to open it, my lungs closed down. Every time my eyes caught its orange spine on the shelf, I gasped and choked in breathlessness. Why had I even been sent this book, and not a monograph on transcendental Thomism? Finally K got the dangerous object out of our house. What she did with it, whether she gave it away or threw it away or read it herself while drinking martinis at some local watering hole while talking herself out of leaving her useless husband, I've never known.

I walked past UVA-CAPS twice a day, the Center for Counseling and Psychological Services. It sat smack between our apartment and the university's classrooms and libraries. I didn't know what lay behind its doors, and the day I traversed its threshold—well, that was an act of desperation. It was also a recognition that the paradigm of Christian self-help, of getting deeper in the Word as a cure for psychic anguish, was exhausted. And it was a recognition that I didn't want to kill myself.

A staff psychiatrist named Dr. Fels happened to be filling hours the day I walked in. She was four or five years older than I, and, in a very soft and reassuring voice, she said some simple things to me—"You're not going to stop breathing. What's happening to you is not unusual. Why don't you tell me more about what you're experiencing?"—and because of those words, I was able to revert to a mean. As I was leaving the clinic, Dr. Fels invited me to come talk to her again.

I saw her a second time, then a third. We began with a weekly appointment, then moved to two days a week. What did Fels do in those hours that helped me so? Perhaps most elementally, she did not seem to find me a freak. At the time, her acceptance of my condition as an ordinary problem she'd seen before and would see again simply felt reassuring, but later I came to understand that her normalizing my condition did a deeper kind of work. There had been something profoundly solipsistic about the formation of my evangelical ego: all the messianic tropes; the sacred only child to my long-suffering, eager parents; the heroic kind of piety. Fels's calm certainty that my anguish was not unique calmed me; it began to erode the narcissism I carried alongside all my fears.

Initially, her questions seemed strange to me. She wanted me to recollect and connect, to notice, for example, whether my present anxiety-inducing circumstances echoed anything from my childhood. Initially, answering such questions felt like a betrayal of family honor, a betrayal of a rigid and right rule of domestic privacy. That I was being invited to talk with honesty about myself, my marriage, my parents, my desires, seemed actually unbelievable to me. But the invitation proved irresistible.

What do you think would happen if you followed through on the conference invitation and went to Amsterdam?

I can't go.

What do you think would happen if you did?

I would be overwhelmed by terrors. And a really bad jet lag that would set it all in motion.

What is it that frightens you the most about the prospect of going to Amsterdam?

If I go, I'll go without K, and if I go without K, I'll be tempted to sleep with someone else.

You fear you'll be unfaithful to K in Amsterdam?

That I'd meet someone at the conference, and we'd have sex.

And then what would happen? What do you think would happen if you had sex with another woman?

I'd go insane.

How would that work?

In a moment of passion, and drunk, or whatever, I'd lose control and have sex, and then after the . . . you know, climax . . . like right after, I'd realize what I'd done. I had destroyed everything—my mind would be ruined—and I would be undone by madness—and my parents, my mother would know too. I mean she'd know what I'd done the first time she saw me back in the States. And I'd have to kill myself.

Why do you think that?

That she'd know?

Yes.

Because my mother can read my thoughts. She can see into my thoughts. I can't tell you how many times my father has told me this.

But parents say all kinds of things to their children. Why do you think your mother has the power to see into your thoughts?

I really needed my mother—it was only her presence that could console me.

If you weren't worried about having sex and going insane at the conference, how would you be thinking about the trip and what you'd like to do?

I'd like to go to Amsterdam and never come back. I'd like to read and write and be free and disappear.

What does it mean to you to say you'd disappear?

I'd like to start over. I wouldn't have to worry about being spied on. I'd feel strong on my own. Strong enough to stay in my skin and not be afraid.

In those first sessions, Fels offered a number of simple reassurances whose emotional impact far outweighed their literal sense. She told me that I was not going to stop breathing, that I could trust my body and its autonomous operations. She told me again that I had not lost my mind, that the avalanche of the disturbing thoughts would not destroy me. We could talk about them one by one. "If you have another breakdown," she said, "it may feel to you like a complete breakdown, but it's not complete. We can do the things we need to restore you."

I would have said—did say, still might—that all those months, all those hours, I could feel these words as bodily salve. I would have said—did say, still might—that these words felt like an answer to prayer. I would also have said—still might— that they were better than prayer, they relieved me from prayer. They relieved me of the burden to turn inner torment into a sacrament. The language of the therapeutic hour felt like prayer reciprocated.

———

Is this what it means to ascend the scale of erotic desire until I find God?

Three or four years in, the marriage felt like a great confusion. Where was the woman I'd fallen in love with—who wore the angel-sleeve paisley dresses, read Gide, and covered Joni Mitchell songs at Cambridge coffeehouses? She was now getting up at the crack of dawn and dressing in modest educator attire to teach elementary school. She was now, in my embittered view, turning into her mother, or mine. In fact, K was doing the same things graduate wives typically did: working thankless jobs to pay for rent and groceries, falling asleep on the couch around the time we'd once met up with friends at bars. I knew K's colleagues and students adored her, but only later would I appreciate the extent of her accomplishments: she developed the first English as a Second Language curriculum in Albemarle County, a sine qua non of the International Rescue Committee's picking Charlottesville as a host city. Meanwhile, I read Derrida and Foucault, and wiled away the hours in my ivory towers.

It's not that I didn't enjoy sex with K. Rather, enjoying sex with K, in our cautiously conjugal maneuvers, made me yearn for decidedly nonconjugal sex with all the women whirling around class, around town, around church. I was shocked by the ferocity of my desire to dive headfirst into the thighs of every interesting woman who caught my attention. I felt the predictable things, trapped and tricked. I fantasized about K's death, my parents' death, my death.

Saving myself for God's chosen mate had not fulfilled the desires of my heart. I was glad to be having sex, I loved the way our bodies felt, but now that I finally was, I began to regret the pleasure I'd faithfully denied myself before marriage, all the unspent passion I was ready to cash in on. Untreated neurotics make great narcissists.

An ordinary afternoon: you might find me in my home office reading, perhaps Schelling, Schopenhauer, or Fichte. I'm taking notes fastidiously—crafting flow charts—meticulously transcribing and outlining. My notes detail the lines of the "I," the all-powerful nineteenth-century German subject that constitutes the world.

But as I read, I'm aware of the windows of my office in the townhouse. I draw down the shades. I switch on the AC unit, not because it's hot, but because I want to insulate the room with white noise. Am I aware that my journey has led to a windowless monad? It's possible that in those years I may have entered a zone between, say, ten and twelve in the morning, during which I could feel that I was capable of being a rational, thinking subject with sufficient resources to understand the world—but that reprieve ended at noon.

I've tried to arrange my life to block out all distractions, but it doesn't take more than the light changing over the afternoon or the perfume in the air when I step outside to arouse countless ancient fears and longings that I don't understand. I might look at a photograph that I took in high school when I was in the photography club and I drove west of town and walked in flat, endless fields of wiregrass; perhaps the photograph reminds me of a time when I felt at home in my body. K is at work teaching ESL to elementary-school kids—and the light or the perfume or the photo turns my mind to porn.

I'd not thought much about porn since junior high school. This was 1984—getting a hold of it required some planning. I might muster up the courage to run out to the magazine stand on the corner, then drive swiftly home and search for the right photograph for quick self-pleasuring, onto the page. That would be followed by a sudden desperate attempt to destroy the evidence, a destruction that would not only guarantee that I wouldn't be found out, but that would, I hoped, mean the event had not actually happened. Then Psalm 51 or a brisk penitential jog through the melancholy woods. Then later, after dinner with K, after K went to bed, I might creep back to my office, often with a beer or two, always with the shades down and the AC roaring, and deep into the night I would read of the ethics of Kant, the powers of the rational soul, the exquisite symmetry between the human will and the divine.

———

One bright and rich April afternoon: Charlottesville seemed warm and soft, but this day was breezy, and the April light here can sometimes have an ethereal hue—a kind of ancient southern melancholy that in some ways may beckon very early memories of my mother's depression in Alabama.

It was after a seminar on Schleiermacher or Coleridge, and Sheila and I walked to Elliewood Avenue, as we often did after a seminar, for a cup of coffee, and then on through the afternoon to her apartment. She lived in an aging hexagon with eighteen- or twenty-foot ceilings, huge bedrooms, and a windowed, corniced living room that had become a salon for gatherings, parties, and dances. She was reading the second volume of Paul Tillich's systematics—in fact, she'd borrowed my copy, which was the hardback my father had read in the late 1950s. On the sofa, we smoked some of her Dunhill cigarettes. If I hadn't been with Sheila in her apartment, I'd have been slashing through another four hundred words on my dissertation (a chapter that at the time I was proud to title "The Overabundant Self and the Transcendental Tradition"), which I was writing for a professor who was inscrutable and distant.

But instead I was at Sheila's apartment, and we were drinking a glass of wine. She talked about her boyfriend, Miles, who lived in Philadelphia and who came to Charlottesville some weekends, when, sometimes, K and I would join the two for drinks. Sheila assumed they'd eventually get married (they did not, as it turns out). We drank more wine and retired from the kitchen to her bedroom—still the sunlight, vertiginous. It was a familiar, ordinary day, except that on this day her mouth found its way to my unzipped jeans.

Was it then or later that I thought of Romano Guardini's discussion of Augustine, in which Guardini speaks of "amazement over existence"? My first response was shock—and then astonishment at this Dutch Reformed woman's gift of tongues—now I knew! And then I thought—maybe, I can't remember—of my sublime Joni, "I could drink a case of you and still be on my feet," and now I knew that too. I had crossed the line and was now somewhere both exhilarating and utterly terrifying—which is to say, led by Sheila and eager to follow to the place that I had always imagined would be the place of my unraveling. What comes after this event?

Sometimes, when Miles came to town, the four of us got together for dinner—Sheila cooked gnocchi, I remember, and asparagus crunchy with salt. Often, I left Sheila's apartment urgent to see Dr. Fels. In her therapist stall, I came to terms with the simple fact that I would survive this. My marriage might not, of course; there might be consequences, but I would survive it. I'd never made a sacramental confession, but even I could see that Fels's offices were both priestly and clinical. What comes after this event? The spring continued, tender and bright.

Cathedral Light

On the terrace of our apartment behind Eddie's Market of Roland Park, K and I drank our gin-and-tonics at the end of the day to the reassuring sounds of the sports fields nearby and our baby's sweet chirps. The double spires of Mary Our Queen reached above the tree line on the eastern horizon. After graduate school, we'd moved to Baltimore with a baby boy in tow. I'd landed a job at a Jesuit university squeezed into a pretty, green urban oasis. The school impressed me with its energetic faculty and a theology department that called itself ecumenical and catholic with a small *c*. I was happy to join the department and, after too long in the valleys of Virginia, happy to move to a city that reminded me of New Orleans.

Still, late mornings I jogged through the leafy streets of the neighborhood dreaming of a different house and different life. I sat at my desk in the evenings after Henry had gone to sleep fairly hyperventilating over the prospect of a report due in a committee meeting the next day, while also trying hard not to think too much about how I had ended up teaching Theology 101 to northeastern Catholic students trying to fulfill course requirements without opening a book.

The sweetness in my life I held gently in my arms, but I could not shake the fear of sudden loss. Is this how it feels to be thrown into existence? A pop song got stuck in my head, and when I heard the rock star sing it at the 9:30 Club on my thirty-third birthday, I wanted nothing but for us to disappear together into my fantastical bungalow at the end of the sea-shelled road. Because, damn, I belonged a long way from here.

I contemplated how I was crazy in love with my blue-eyed son. I desperately wanted to be the father he deserved, to be present and steady, not felled by the symptoms that lurked behind the shadows drawing nigh.

———

K and I joined a church in an inner-city neighborhood. The congregation comprised of elderly holdovers from the church's glory days and a band of young adults inspired by a recent movement of left-leaning evangelicals worried about rich Christians in an age of hunger. The year was 1991, the third year of the first Bush administration, around the time the Gulf War commenced. I lamented Billy Graham's visit to the White House, where he prayed with the president not twenty-four hours before Bush declared war—and was delighted to find a church that shared my dovish tendencies. These good souls were Dorothy Day people; they were feed-the-poor, protest-the-war people. They were "Three R's" people, to recall John M. Perkins's social-gospel summons: "relocation, redistribution, reconciliation."

I was happy to number among the fold. I thought that, in

these good people's company, I could redress some of the sins of my childhood, and I thought I had found a way of being faithful that would allow me to breathe. I'd turn my spiritual striving toward the neighbor and immerse myself in deep community, thereby sidestepping the pious obligation to introspect. All the energies I'd once unleashed on scrutinizing my own waywardness I could now put to the task of scrutinizing society's ills and healing them, in the name of Jesus.

Building community with indigents and immigrants, forging biracial partnerships with BUILD, IAF, Habitat, and Catholic Worker, we aimed our sights on the broken social order; we were coalition kings for the kingdom long before intersectionality was cool. With enough resolve and discipline, Old Seventh Baptist Church would flourish in the direction of our social hopes, and we could serve without noblesse oblige the minority families living in the decrepit row houses nearby. We would bear witness to forgiveness and reconciliation in the spirit of lament, which meant drafting a confession for the sins of whiteness. It meant gathering each week in members' homes over lentil soup, day-old bread, and readings of the Minor Prophets to pray for our demonstration plot for the kingdom at the corner of St. Paul Street and Seventh Avenue.

Nothing worked according to plan. On the rare occasion one or two Black people from the hood might drop in on a Sunday, the gassy ardor inside would knock them back on their heels and down the front steps, never to reappear. On Sunday mornings, we fought back discouragement as we hurried from our cars to the battered sanctuary, our precious mimeographs floating down St. Paul like sagebrush in the desert.

Our minister was a kind, soft-spoken man whose wife taught special education in the city schools. On potluck nights, he played Crosby, Stills, and Nash songs on his acoustic guitar. He told stories of his Kerouac-inspired road trips to the West Coast. But one Sunday morning he made a comment in a sermon, little more than a passing remark really, that gripped me with a panic and made me begin looking for an exit strategy. The pastor asked us why we had decided to follow Jesus into this blighted soulscape—fivescore souls huddled against the cold in a sanctuary once warmed (before white flight) by seven hundred bodies every Sunday. Why had we joined this bold venture of faith?

Because, he said, the Christian life was the most intense life of all. "There's nothing more intense than following Jesus, man."

And it was precisely then, on a blue winter morning in my thirty-fourth year to heaven, that I realized I was done. I could live with the peeling paint, musky pews, and dust mite–filled runners—and even with the twilight glow of the chandeliers high above and the invading torpor. What I could not handle was the prospect of greater individual intensity. I lacked the spiritual verve and inner resources or whatever it was that I had long leaned on in my attempts to conjure purpose out of solitude—the evangelical ambition that grinds away the spirit like bone on bone.

I'd reached the end of a long road. I had nothing more to give. I had hoped that life among "renegade Baptists"* would be

---

* The phrase is Will Campbell's—a self-description.

free of imperatives about intensity—but that hope, of course, was mistaken. To be sure, many devotees of the social gospel are driven as much by love of Jesus as by neurosis, but take an overly scrupulous, anxiety-broken only child and place him in a soup kitchen on a street of boarded-up row houses, and what you get is a man who experiences, in the quest for beloved community, merely a different aspect of the weight of righteousness.

Call it a crosswise altar call. The pastor's words that morning set in motion a sequence of events that changed my life.

———

Why therapy in Baltimore? Work and love, as Freud said. I didn't know how I'd ended up teaching theology at a college I'd never heard of until its name appeared in the AAR job listings. My wife seemed oblivious to my suffering. My sufferings made me oblivious to hers. Our marriage had barely survived graduate school. If she only knew how much I wanted another life; if she only knew how elementally distracted I sometimes felt in her presence. Had I not held such an apocalyptic view of divorce, it surely would have ended. But we held on to each other like "children in a circle dancing"—that line from W. H. Auden inscribed in my wedding band—and now we had a baby boy to deflect our evasions. And K had her own demanding parents to deal with, or to not deal with, the approach she and her four siblings preferred.

My time with Dr. Fels had given me tentative, temporary respite, but now there was a chasm. Or maybe now the stakes were simply higher—what I wanted, more than I wanted to be a decent husband or a scholar, was to be a father to my son. And now, leaving Old Seventh, I no longer had the patina of purported godliness to pretty up the ways I was, I feared, already failing at that task, at that calling.

I needed help. I wanted to gain control over the goings and comings of disturbing thoughts. With the aid of that indispensable guide to the outside world, the Yellow Pages, I found a psychotherapist near my college and began seeing her once a week. Talking with a skilled and empathetic listener, even only weekly, made me feel hopeful and my symptoms seem less absolute—I felt less confined to them and alone—and that was no small accomplishment.

Still, after a while my therapist recommended psycho-analysis. I didn't take kindly to her pitch. I was an assistant professor making less than $26,000 a year. My wife was at home with our three-year-old son, with another baby on the way. By my calculation, analysis would cost more than half my annual salary, about 70 percent of my take-home pay. The therapist regarded my concerns about money with the suspicion native to her guild—surely I was using finances as a feint. In commending analysis, the therapist surmised that I lacked the repertoire to understand the particularities of my suffering. I'd seen her certificate from the Baltimore-Washington Psychoanalytic Institute—next to her MSW and her Mondrian prints—and early on sensed she wanted to get me into the whole Freudian harmonium. I walked out of the office and never came back. Nor did I reply to her handwritten note asking if we could talk.

She was right, of course. Later, on an evening not long ago, I googled her whereabouts and was surprised to learn not only that she was the daughter of a New York art collector, but that she'd worked as a field organizer for a poverty-relief agency in Appalachia and remained active in progressive causes until the night she was hit by a car crossing Boxhill Lane after book club. Seeing her photograph in the *Baltimore Sun* filled me with sudden gratitude for the year during which our sessions kept me steady and steered my thoughts toward analysis, and with regret that I'd lacked the repertoire to thank her.

My departure had little to do with her candor. You could have filled a football stadium with the people who had a plan for my life, people who wanted more, more, more. The last thing I needed was another. Still, the fact that I left more hurt than angry meant, as I thought about it over the next few weeks, that I did not entirely disagree with her proposal.

I'd taught Freud in my courses on modern religious thought and lectured as a graduate student on his explosive little book *The Future of an Illusion.* While traveling in Vienna for dissertation research, I'd taken advantage of a multihour layover and caught a streetcar from Laxenburger Strasse to Berggasse 19, to the unassuming building where Freud lived, kept his clinic, and held his weekly seminar. Then I walked to the university and paid homage to the black marble bust of Freud in the courtyard.

In recent months, I'd compiled a small library—on a shelf

over the doorway of the sleeping porch—that included Erik Erickson, Karen Horney, D. W. Winnicott, and Jonathan Lear. I can't remember how I learned of the Yale psychoanalyst and professor of psychiatry Stanley Leavy. His name would have surfaced in a search for "psychoanalysis and religion"—among the results that were not conspicuously tendentious. *The Psychoanalytic Dialogue* and *In the Image of God*, on the career of an analyst, and a Christian one at that, came to me as a gift.

But it was an essay by a Baltimore psychologist named John Gartner, in an edited volume on religion and mental health, that would eventually lead me to the office of an analyst named David Lieber. The article weighed in at under eight pages of brittle prose. "Religion has the potential either to increase or to decrease anxiety," read one of the article's key insights. It's hard to disagree with that. Yet I was excited by the notion that a psychologist working nearby pondered "the cry of anxiety-driven loneliness" and "the distortion of the God-idea with the fierce super-ego." I left a long message on Gartner's answering machine that must have intrigued or worried him. In a matter of days I was in his office, admiring his Eames lounge chair and ottoman and the wall of posh diplomas in matching frames behind him.

When Gartner asked me what was on my mind, I picked up where I'd left off in my voice mail, except this time I told the story of my nervous illness in an artfully crafted fifteen-minute narrative that I'd written out in longhand and rehearsed on the walk through the leafy grounds of Sheppard Pratt Hospital, where he kept clinical hours some days. When I was finished, he told me about a young analyst in training, a "brilliant man," he said, who was looking for a willing analysand. He pulled a sheet of paper out of his drawer, jotted down a few words, and recommended I read Lieber's recent article on Kafka and the unconscious.

"This is an extraordinary opportunity, Charles," he said. "If you're interested in moving forward, I'll let David know we talked."

As I wandered back to the parking lot, I opened the note to see what Gartner had written. On the otherwise empty sheet of watermarked stationery was Lieber's phone number on one line and on the other: "5/11/94. 45 minutes. $120." And that was the last I heard of Gartner until, years later, I read in a newspaper that he'd defied the American Psychology Association and its diagnostic protocols by declaring *ex cathedra* that Donald Trump was mentally ill and unfit for office.

———

Throughout two masters degrees and a doctorate in theology, I had tried to align my mental illness with the Christian mandate. I'd prayed for the grace to persevere, to love the torment, to receive the martyr's patience. I'd prayed for healing and the promise of new life. Even as my soul, my irrepressible, God-formed somebodyness, pushed back.

And then Lieber. Call it the comedy of grace, my dogged persistence, or the infusion of both: my prayers for healing were answered in the form of a balding Jewish analyst and MD who offered an excellent sliding scale. Lieber explained the basics. A complete analysis would take about three years. We would meet four days a week. The hospital's usual rate was $150 for a therapeutic hour, but since this would be a training case, which he needed to fulfill a clinical requirement, he'd be willing to accept less. I should think about what I'd be willing to pay, a rate that seemed feasible. He would not be earning any additional income on the analysis.

A week later, I explained about the $26,000 salary, that my wife had taken leave of her ESL position to care for our two small children, and we had recently bought a two-bedroom, one-bath house, for which we'd borrowed heavily. I was thinking I could pay five dollars a session.

Lieber said that sounded fine. He didn't seem taken aback by the amount; he said only that he thought we might agree to revisit the rates, should my financial outlook improve.

During the last of our three initial consultations, I told Lieber I appreciated his offer—and his patience with my questions—could I have a little more time? Three years of near daily therapy seemed daunting, even at bargain-basement prices. Lieber said he understood. I was to take as much time as I needed. The whole thing felt thrilling—equal parts danger and oh-hell-yeah.

Some nights I woke up thinking about the implausibility of the whole thing. To get to appointments on time, I'd have to leave at the end of my second class, no hanging around to answer questions or shoot the breeze, walk to my car, hope the car had not been stolen or broken into, and drive the speed limit from Roland Park into the urban desolation of east Baltimore—set of the popular television series *Homicide*—gunning through yellow lights, as I headed toward the massive concrete and steel oasis of Johns Hopkins Hospital. But that wasn't all. Once there, I had to find a parking place, hopefully one in the Rutland Street garage nearest the psych building, navigate a labyrinth of hallways and elevators to the tenth floor of the Meyer Wing, where finally I'd enter a crowded waiting room and sit until Lieber's door cracked open. And then fifty minutes later, I'd have do it all again in reverse to make my afternoon class. And this would be my life, four days a week, with breaks at Christmas and in August, for the foreseeable future.

Sometime during those weeks, the psychiatrist and writer Robert Coles came to Baltimore to lecture. It was a greatest hits performance—Flannery O'Connor, Thomas Merton, Walker Percy, and as always Ruby Bridges—and although some of my colleagues were annoyed he'd not given the audience anything new ("Ruby Bridges again. Really?"), I was ecstatic. They were happy to accept when I offered to take Coles to breakfast the next day and then on to the airport.

Because I'd long been an admirer—I'd especially adored his account of school desegregation in New Orleans and *Walker Percy: An American Search*—I took it as a sign of providence that our paths were crossing just now. At a bakery in Roland Park, I steadied my pen and opened my notebook. There were many things I wanted to ask him about. Walker Percy, the writing life, and civil rights, but mostly about psychoanalysis and whether he had any advice for me. Of course, he did.

"Why would you not say yes?" he asked. "I'm about to go into my third round."

———

Why wouldn't I? Understanding and healing had been the one consistent thread in my journals and notebooks. But an evangelical in psychoanalysis seemed a contradiction in terms. Evangelicals might do pastoral counseling, which could mean listening to your pastor ramble about sin and how, he is convinced, most mental problems are the result of disobedience to God. You might then be encouraged to take a cleansing dive into the Psalms and to check back a month later.

In one of the few surveys on Christianity and mental health, LifeWay Research, of the Southern Baptist Convention, concluded that 48 percent of self-identified evangelicals said "they believe that conditions like bipolar disorder and schizophrenia can be treated with prayer alone."

Psychotherapy, on the other hand, with its origins in the Greek words *psyche* and *therapeia* (and the movie *Psycho*), loosened the fears that protected you from the world while conjuring the grim prospects of a grown man chatting with his dead mother in a room full of antiques. Anyone saved by the blood of Jesus said no to psychotherapy and most certainly did not get on the couch.

For true Christians—not the reprobate moderns with their higher criticism and milquetoast religiosity—repression and guilt remain a bulwark against sin and collapse. Repression and guilt are in fact the true Christian's allies. The theory of spiritual warfare is the only hypothesis capable of explaining inner torment. In my notes from my father's Sunday evening sermon, September 24, 1972—notes that I gave the title "The Work of the Devil"—three arrows follow the words "Satan's Main Instrument" pointing directly to "ME."

A book titled *Competent to Counsel,* by the Christian psychologist Jay Adams, gave credibility to many evangelicals' native distrust of psychology and psychiatry. Adams claimed that a few mental illnesses were caused by problems like a brain tumor, but most people claiming mental-health problems simply refused to meet head-on the standard-issue challenges of life; any pastor familiar with the Bible was "competent to counsel." *Competent to Counsel,* which reads like a "too blessed to be stressed" manifesto, inspired the Bible-based counseling movement and continues to influence evangelical approaches to mental illness. In a companion volume, Adams wrote:

> Biblically, there is no warrant for acknowledging the existence of a separate and distinct discipline called psychiatry. There are, in the Scriptures, only three specified sources of personal problems in living: demonic activity (principally possession), personal sin, and organic illness. . . . All options are covered under these heads, leaving no room for non-organic mental illness. There is, therefore, no place in the biblical scheme for the psychiatrist as a separate practitioner.

As if to "fight Freud with Freud," Adams included in *Competent to Counsel* his own case studies, for example, of a lazy college student who faked nervous breakdowns to avoid final exams and an adulteress whose manic-depressive behavior continued because she was indulged by counselors (she started to be cured the moment she received, in mid-hysterics, a stern reprimand: "O be quiet! Unless you stop this kind of nonsense and get down to business, we simply can't help you, Mary").

Let's summarize it this way. At the time of my breakdown, the evangelical subculture remained largely committed to a sin-based explanation of mental illness. That commitment was only secondarily a view about mental health; it was first a wedge in the separation of evangelicals from liberal Protestants, who in the nineteenth century had begun talking about synergies and correlations between Christian belief and modern views of the self; in turn, ministers and laypeople aligned with the social gospel began creating asylums and psychiatric hospitals. Evangelicalism ossified in opposition to liberal Protestantism and its alleged concessions to the scientific worldview and promiscuous use of secular psychologies.

All of which to say, maybe I wasn't an evangelical any longer. I phoned Lieber to tell him I was in.

———

On a summer afternoon in 1994, a thirty-six-year-old anxious riddle of a man got on the couch and begin talking.

"What do I talk about?" I asked.

"Whatever you want," Lieber said.

"That's it?" I asked.

"Pretty much," he said.

You might say an evangelical childhood brings certain advantages to the work. I'd learned early on to share my faith with anyone interested (or not), to offer an account, on demand, of my spiritual life. When you give your testimony, you are free to talk about anything. Intimate matters that came between you and God, whatever weighed you down. Your troubles, trials, and secret sins, the things that kept you from being all you could be in Jesus. I confessed to a room of well-fed Baptists that I had found a centerfold in the woods behind our house and looked at it—there always seemed to be porn in the woods. I confessed to this and more, and sometimes, because I lacked a dynamic testimony, I made things up. Analysis felt like a very long testimonial minus the *amen*s and *praise Jesus*es.

I told Lieber I wanted to talk about heaven. An image came to me from Jesus camp. I couldn't remember which camp. Jesus camps loomed over every sad body of water between Picayune and Dothan, and I'd passed summers at so many. It might have been the retreat on the Gulf, or the ranch in the east Mississippi lake country, or a weekend bivouac on the banks of the Tombigbee River. But I recalled vividly how on an afternoon swim I had positioned myself underwater by the ladder, with goggles and snorkel, and watched the older girls lapping in the deep end overhead. How their bodies flashed like pearls. How I followed their motions back and forth in the aqueous light. I was hard enough to hold a tray of oysters, but no one could see me underwater—no one paid attention to me. I was the preacher's kid splashing around in his own little world. Later

---

Notably, the Magnolia State consumes more porn, which includes gay porn, than any other state in the Union (with Alabama and Louisiana a close second and third).

that evening at the worship service, I sat in a folding chair and listened to my father deliver one of his powerful sermons, and through a window open to the fragrant night I watched the last light of day smear the sky like lipstick.

During another session, I steer my thoughts once again to the Gulf. Calm waters lather the shore. The sky shines radiant and blue. I truck to the crest of a dune and lie motionless on my back. I think at first it's a sea bird's caw, but soon recognize a familiar song. From the campgrounds, the girls walk, arms linked, toward the beach. "Two men walking up a hill, one disappears and one left standing still," they sing. "I wish we'd all been ready."

I turn to watch the girls moving slowly across a warm tongue of sand in their cutoff jeans and tank tops and hum along to this sweet dreamy ballad of guns and war and total global apocalypse. "There's no time to change your mind, you've been left behind." Forever bound to this coastal high-way, and bodies liquid in sunlight, and the thrill of immersion, and the awful fact of hell.

Other days, I spoke of the shape of the bedposts of my child-hood bed, the heaviness of my mother's nightdress, the hair on my father's hands—the archaeology of my fears. I spoke of my desire to extricate myself from K and Henry and Will, and my terror of extricating myself from K and Henry and Will.

"Does that give you something to work with?"

Lieber responded with a gentle "Uh-huh." The first of a million such *un-huh*s and *hmm*s I would hear in the coming years. A reassuring sound, I felt, signifying a thought well spoken, an insight gleaned. The universal signal for progress in therapy? Or nothing more than the sound of the feigned interest you might make during a committee meeting while increasing an eBay bid on your laptop? Who could know for sure?

Lieber, having noticed how asides about the apocalypse dotted my memories, said he'd like to hear also about my earliest thoughts of hell.

Okay, that was a total buzzkill. But since you asked, hell is a place of repetitive torments custom-made for the vagaries of the singular man. Birds pecking at the sides of the damned hanging by their hair over bottomless gorges, menacing footsteps shuffling in the hall, gargled breathing, a pounding heart, the growls of lunatics. Alone in the park that grows predatory at sundown as twilight dims the day. Like St. Peter, who sees the angel Ezrael bringing children down to Hades to see how the disobedient are punished. Aborted fetuses shooting fiery arrows into the eyes of their mothers. The gnashing of teeth, the smacking of lips—remember how crazy it would make you to hear your mother's chewings?

Later I'd read that hell is falling into the place "where the fear is." The anxiety that Jesus will return and you'll be left behind and suffer eternal damnation creates a special kind of anticipatory fear. It is one that endures long after the doctrine has collapsed into doubt or disbelief. My devotion to the suffering Jesus had bled into the bliss of a deep, warm thigh. The more my gratitude swelled for the blood shed for me at Calvary, the greater my howling desire for a girl's soft summery body. Faith was fire, and fire was sex, and sex was death . . . as the seasons go round and round.

———

Psychoanalysis would not lead me into paradise, and what began as a joyride along the lost libidinous highways of repressed desire settled soon enough into something that felt more like an ascetic discipline.

Analysis and faith traverse similar terrain—they understand how language and narrative heal. They may see each other as strangers or competitors, but they need not. Like prayer, the analytic dialogue slows down to ponder, to meander, to piece together, to redeem; both inspire the mind toward hope under the influence of an empathetic listener. Neither needs the other to effectuate its truths, but they follow parallel tracks into the mysteries of being human, where all truth is God's truth. It's more than fine that they neither merge nor collide.

It seems to me now, looking back, that analysis helped restore vitality to a prayer life grown predictable and palled. "The God who calmed the seas can calm my soul," I wrote in my journal, after some months with Lieber. I meant, surely, that God could use analysis to confect that calm. "On this clear bright morning in early October, I pause to offer my life to God's great joy. Make my life a joy," I wrote. "Keep me safe, O God," I spoke, with the psalmist.

Which is not to say that gone were thoughts of my own unworth: "Repent, then, and turn to God, so that your sins may be wiped out," I quoted from Acts—but not, "The wedding is ready, but those who were invited were not worthy," and not "Every tree that does not produce good fruit will be cut down and thrown into the fire." Rather, I chose Acts' instructions to repent—"that times of refreshing may come from the Lord." Those times of refreshing seemed possible, and it seemed possible that I would know their source.

I want to make a case for the theological integrity of psychoanalysis. A skilled analyst—one who, like Lieber, has a sensitivity to religious experience—can help the analysand bring trauma and wounds and joys and, for me, anxieties and inexplicable symptoms, into an incarnational ordering of language. The miracle is the way language can reconcile disparate parts of the self. Freud's great discovery was identifying the psychic mechanisms of idolatry. The patient's sick ego promises us candor, he said—candor that is unavoidable, given the material that the analysand's self-perception provides—the terrors, the desires, the inhibitions.

In its strenuous exhilarating exertions, I experienced a second childhood. My thoughts rushed with sudden intensity back to my earliest fears as I wandered "spellbound . . . through a forest of memories."

The immersive wordiness of analysis and the worldliness of the Incarnation—"The Word became flesh and made his dwelling among them"—conjoin in their affirmation of freedom. What does it mean to be free in Christ? It means the capacity to name the false gods and the power to resist them; it means defying the tyrannical voices that Christians so often endow with a kind of religious aurality. And, for me, it meant hacking my way through a magnolia jungle of wounded memory. The psychologist who'd directed me to Lieber had spoken of analysis as an optimal narrative of the person's unique becoming, which seemed about right.

Since I first read his 1943 letter from Tegel prison, I was gripped by Dietrich Bonhoeffer's use of the musical image of polyphony (that style of composition in which multiple melodies somehow cohere into a whole) as the pattern for life abundant. For Bonhoeffer (who was not uncritical of psychoanalysis), polyphony becomes the last and satisfying metaphor of Christian life—Christ as *cantus firmus*, the melody upon which the counterpoint is fastened. The polyphonic lines can then reach into the extremes and can go where they will, as they are then centered by the *cantus firmus*. Bonhoeffer wouldn't say this, but his metaphor is, in some respects, also an image of analysis. Analysis is the space where one feels—where I felt in an embodied way, in the unhurried hours over months and years—a trust in the beautiful interplay between the center and the extremes.

My body and mind would not be raised in resurrected splendor in the course of the treatment. I wish to emphasize the point. It was tempting to think that it would, that I would undergo a miraculous transformation. If not resurrected splendor, then surely I would take on the "new man." Instead, I received the gift of mortal life: the freedom to be imperfect, to have fears and face them, to accept brokenness, to let go of the will to control all outcomes.

Some sessions left me feeling oceanic and unbridled; others ground away in shadowy stasis like some purgatorial treadmill. All part of the work, Lieber would say, raking his fingernails across the stubble of an early morning shadow. Rarely, though, did I feel that I was wasting time. The unhurried back and forth, day after day, the fits and starts, and the awkward silences all became places of illumination and discovery.

I was free to talk about the window smashed in my face by a drunk outside the Buck 49 Steak House in New Orleans; the terror of learning that my tongue was too large for my mouth; difficulty breathing; asthma and allergies; walking and legs; the beauty of my calves and the sin of pride; hearing things— loud and threatening noises, the gollar of unfamiliar voices—as a child at Billy Donlon's house; vigilance in silence—in the absence of God's voice hearing dripping faucets, barking dogs, the mumbled voices of my parents in their bedroom; sleigh bells on the roof; the great hissing silences of the house; "Listen, pilgrim . . ."; sudden hyper body discomfort and practices of self-renunciation; acute fear of confined spaces; night terrors; fears of being abandoned to predatory men.

In the languorous flow of thought and speech, in the recollection of known terrors and forgotten buried ones, of visceral traumas and hopes, under the skillful direction of an analyst, psychoanalysis did loosen me from so many tyrannies of self-unknowing, from cruel dogmas and debilitating fears. In analysis, I found the freedom to feel in recollection, to follow the lines of repressed desires—and the lightning didn't strike. In a slow turning, I learned to trust, for the first time, in the aptitudes of bodily life and to open doors long locked by fears. To imagine a life that no longer means in every instance self-sacrifice.

Sometimes I worried that I had balked at the dark night of the soul—and just before the moment when I would have been transformed. Or that I had abandoned the narrow way for the comforts of the couch. But the daily grind of analysis requires discipline and sacrifice, of course, and becomes its own cross to bear.

———

It was not only psychoanalysis that I found in Baltimore. Some months after leaving the Baptists, K and I drove our children in the beige minivan to Baltimore's Cathedral of the Incarnation, between St. Paul and Charles Streets. It wasn't the first time either of us had worshipped in Episcopal pews. Both Gordon College and K's alma mater, Wheaton College, numbered among their faculty members professors who had discovered Anglican liturgy, fanboys of their worldlier British counterparts; near enough each college's campus was a parish that came to be known by the clunky moniker "evangelical Episcopal church." (A few years after we graduated, one of my professors—a flamboyant scion of American evangelicalism who initiated many a young man raised on Josh McDowell, Maranatha Music, and the Living Bible into the rites of cocktails and the Book of Common Prayer—published a book whose title summarized the curiosity about and call to liturgy on campuses like ours: *Evangelical Is Not Enough*.)

Even before we began dating, K and I had separately found ourselves beckoned by the Episcopal Church's unfamiliar habits: reciting the Nicene Creed as part of Sunday worship, celebrating the Lord's Supper—now only ever called the Eucharist—every week rather than monthly or quarterly. And, separately though virtually in unison, our parents had balked. Though our fathers differed in their theologies of baptism and election, they were of one voice in their objections to Canterbury. Chiefly, their criticism was that the Christianity of Episcopalians was too easy, too comfortable, too smooth and serene. Salvation by good taste alone? The perfect synecdoche of that comfort was the Episcopal Church's use of wine rather than unfermented grape juice at Communion.

K's father pressed us both, on the eve of our engagement, to sign a contract, a covenant I think it was called, pledging our lives to the family's generations-long abstinence from all alcohol, a practice my future in-laws—as well, they claimed, their own parents—had adopted not for the protofeminist reasons voiced by female temperance advocates, viz., that husbands and fathers spent their paychecks on booze and came home most nights in a storm of violence, but rather as an efficient encapsulation of superior virtue. Drink belonged to the likes of liberals, wayward youth, and flight attendants—that is, to the lost. We declined to sign, which ensured our fate as mongrels among purebreds.

Upon marrying, K and I had at first hewed to the denominations of our childhoods—we worshipped, almost exclusively, at Baptist and Presbyterian churches and told ourselves we'd find a home by the time we had children. But upon leaving Old Seventh, we realized that we could no longer locate ourselves in the preaching or prayers of either of the denominations in whose churches we'd been reared. K was tired of hearing Baptist pianists bang out "I Come to the Garden Alone," and I'd had enough of five-point Calvinism (I thought, as I continue to think, that the doctrine of limited atonement is an abomination and an offense—but I digress).

The cathedral's dean eventually invited me to preach, but not in our first year there. He saw me not first as a professor, but as someone who needed a church community with a lot more free space than the ones to which he had been accustomed. On Sunday morning, I was no more or less special than the people reeling from alcohol and homelessness who drifted in for worship; than the Liberians who'd fled the regime of Charles Taylor, bringing with them dance performances that were some of the purest praise I'd ever encountered; than the students who walked over from Hopkins; than the mother of two who came to church only occasionally, wearing heavy makeup—so sheltered was I that it wasn't until she, with the dean's help, had extricated herself that I understood she was married to an abuser.

K and I and the children were—like all other low-church refugees—enthralled by the sensual aspects of Sunday worship at the cathedral—the light, the colors of the stained-glass windows, the spiced smell of incense, the polyphonic chants of the choir, the golden chalice, the vestments. I remember tiny Will, two years old when we found the cathedral, chirping like a bird during all the hymns. Worship was intoxicating for the children, and for K and me as well; indeed, so stunning was the sensorium of Sunday services that I never stopped to ponder the contrast between this bodily and beautifying worship and the stripped down, wordful worship I'd known when I was my sons' age. I was, simply, part of a church at prayer.

It was only after several months of cathedral Sundays that K and I could begin to name to one another how eager we both were to be released from the heavy weight of the Protestant evangelical subject; how desirous of something beyond a worship service built entirely around preaching, performed in a sanctuary bereft of images, iconography, crucifixes, beauty;

how spiritually depleted we'd become from the expectation that, in those melancholy offices, we ought hear the lacerating word, ought hear God speak directly to us, ought somehow find, in that hour of biblical exposition and a few grudging hymns, nourishment of body and mind. Some days, I left the cathedral wondering if, possibly, the place housed too much beauty, too much mystery, too much love.

———

Mondays returned me to analysis. What untold secret would I flee from today?

I begin. Lieber's taking notes. (Or is he? He and his metal chair remain out of sight, according to the protocols of the treatment.) I am again explaining what has brought me to his office. I'm describing the symptoms that continue to plague me. I'm casting about for a history of them, or for an account of the way the symptoms have blocked my capacity to think, to probe, to question. What's the mechanism by which analysis works? How does the unconscious life become accessible through dialogue? The mechanism is language. The mechanism is giving words to ancient fears and inhibitions that, before Lieber's office, outside Lieber's office, had been available only through the body and the sensate.

One day I thought, *This hour of meandering, confessional, soul talk, whatever feels familiar: it is the flesh made word.*

Sometimes in analysis I would describe—and sometimes I would begin again to feel—the anxieties my childhood body had carried. I would tell Lieber about the suffocating asthma, about the fears of being buried alive, abandoned. I would tell him that, as a boy, I wanted the conviviality of a friend—I wanted, I once said to Lieber, nonaloneness. But I didn't have the conviviality of a friend, not really. I was a boy who was afraid of the dangers of the night who could only be consoled by his mother, her warm body and her abiding protection and reassurance.

And I would talk to him about the anxieties of living as a white preacher's child in the Jim Crow South during the most tumultuous years of Mississippi Burning, which is to say I talked about white violence and the threat of white violence. And sometimes, in Lieber's office, I could name how those two lines—my suffocation and yearning for my mother, and my fears (and my mother's fears) of predacious white men—coalesced at the site of my first and most acute anxiety as a child, episodes in which my personal and social worlds clashed and thunder-clapped in my body, in my hands.

A neighborhood on lockdown, several times a week—all lights turned off on my block and the next block and the next block, as though it were London during the Blitz, because all the householders feared marauding groups of men (Black men or white men; it barely mattered). And how I felt the anxiety of my neighbors, and anxiety about my neighbors, in my parents' own jitteriness, a jitteriness that cut off their capacity to comfort me. They tried to maintain equanimity and to comport themselves in a way that would befit a refined, educated woman from Jackson and her ambitious pastor husband, but no one had to explain to me that my parents' big hearts, my parents' Christian love, was constrained by the color wash of racism and violence.

No one had to explain to me that churches were being bombed. And no one had to explain to me the night, the palpable sense of menace; even a child could infer the menace from the fact that all the houses on the street had dogs, that although no one could walk the streets without being seen by all the neighbors, we nonetheless decked our windows and doors with burglar bars and our lawns and windows and porches and treehouses with surveillance gear.

I would describe those things, sometimes, to Lieber.

Psychoanalysis is not just about charting genealogies, unearthing artifacts, or turning over causal loam. Analysis was the creation of a space where I could revisit those places—the defended lawns, the room in which my mother's comfort was sometimes hedged by her own terrors. In analysis, I could be in that room and once again feel the terrors. Why did I want my mother? What would have happened to me if she had not come, had not been there?

In analysis, you can see the child and grieve. You can open the doors, see him, and grieve.

———

Freud called them "psychic intensities": those things—those primal scenes, those deep terrors—that become your regulative pattern. For me, of course, that scene was the possibility that my mother would not come. You feel that scene; you encounter it in all kinds of places. You're at work, and someone says something seemingly innocent or looks at you absently, head tilted, with a certain lilt in their voice, and you are a child, struggling for breath, waiting for your mother to come. Or you're at a Christmas party and suddenly you're not; suddenly you're in your childhood terror. Which means, among other things, that you overload the smallest conversation with a colleague, neighbor, friend, or dental hygienist with an abundance of meaning, so much more meaning than the conversation deserves or can bear, so that it isn't about the current conversation, but is in fact about waiting for your mother—and you can see the confusion blooming on your interlocutor's face.

Your primal scene presents itself before your analyst too. Lieber becomes your mother. You're waiting for her. You're terrified. Over and over you replay this with him, feeling you have no choice but terror—until one day you do have a choice.

Paul Ricoeur, four decades after the publication of Freud's *The Future of an Illusion*, argued in an unwieldly, largely forgotten monograph that Freudian psychoanalysis prepares the mind for a faith cleansed of idolatry—to apprehend the God beyond god. "The question remains open for everyone," Ricoeur said, "whether the destruction of idols is without remainder." The analytic dialogue exacts a meticulous, demanding, and expensive process (unless you have a sliding scale) of disentangling the reality from the symbol, of freeing the transcendent mystery from the domesticated word.

Freud may have exaggerated his conclusions, presuming he'd exposed faith's essential naivete as the natural history of an infantile obsession. You see aspects of caricature in his genealogy of religion, his simplistic reductions and quest for a theory of everything, his confidence in the morality of science—he was a child of the nineteenth century.

Still, Freud's critique of the idea of God strikes me as less a "funerary sermon on religious culture," as one critic surmised, than a reckoning with the seductive power of illusion. Analysis offers a clinical procedure for disentangling symbolic conceptions of God from the reality symbolized. Such at least was my experience.

St. Paul's prayer that Christians would be strengthened in their "inner being with power through his Spirit" and the psalmist's search for "truth in the inward being" and "wisdom in my secret heart" affirm the habits of self-care and depth. St. Peter tells the resident aliens scattered throughout Pontus, Galatia, Cappadocia, Asia, and Bithynia to "adorn their inner selves with the unfading beauty of a gentle and quiet spirit." In the modern age, the theologian Paul Tillich spoke of the *courage to be*, "the power of creating beyond oneself without losing oneself" to enter into the fullness of life.

Our lives are a marvelous mystery.

In Baltimore, I found that mystery first in the kind preaching and communion of a church that wished not to produce affect, but rather to accompany with love—and that mystery began to return K and me to a gospel I had half begun to suspect lived not at all under the promises of our childhood pieties.

Sometimes, at the weekday services of Morning Prayer, I'd be handed a Bible and asked to read the lesson—a reading from Ruth, a reading from Ephesians. Once, the appointed text was from Revelation, every evangelical child's favorite doom scroll. But now:

> At once I was in the spirit, and there in heaven stood a throne, with one seated on the throne! And the one seated there looks like jasper and carnelian, and around the throne is a rainbow that looks like an emerald.

When I closed the Bible, I saw the translucent morning light arcing through the stained-glass windows. The light brought Revelation closer to me, and me to it.

And I found the mystery a second time too. Under the direction of a competent analyst, psychoanalysis builds upon the mystery in its respectful listening to the person—to the suffering self, the analysand. Healing began in the unscripted exchanges and transactions of the analytic dialogue, in the silences of the hours.

Outtakes from an
Evangelical Analysis

V erbs for my time with Lieber: *unspool, unwind.* I unwound,
or Lieber's questions unwound from me, stories, as if on
an old cinema reel.

I think of my mother making taffy. The verb for making
taffy is *to pull.* Mother would pull the mess of boiled sugar and
butter, pull and pull and pull. Or another verb, *aerate.* The key
in the art and science of taffy making is introducing air into the
sugar and butter, small bubbles of air, small pockets of breath.
It's the air that gives taffy its lightness, its texture.

Sometimes, when I am thinking of my time with Lieber, an
image of Mother pulling taffy comes to mind.

I

On the couch—the proverbial couch, which was really a daybed, but which looked frankly like a military cot—I told some
stories over and over, pulled this way and that.

I told about the early summer evening my mother and I
boarded a new Continental jet for Houston. My father was

preaching a weeklong revival for an old seminary classmate who'd been such a close friend that they'd stood for one another at their respective weddings.

In Jackson, my mom and I settled into comfy first-class seats for the short flight. She sat with her Bible open, while I admired the gold-tinted seats, paneled interior, and soft lighting. After takeoff, the airplane cleared a heavy cloud bank over the Mississippi River into a sunset unlike any I had ever seen. My mother took my hand as the sky turned orange and red.

Andy's wife met us at our gate with a bouquet of moss roses for my mother and a new football for me—the detail of those roses surfaced maybe the fourth time I told Lieber about this trip. Mrs. Hanks was even nicer than I had expected. She touched my hand lightly with hers and patted my knee as she drove to the parsonage. As the car moved through the thick, sweet heat, she asked my mother to pray for my young life, that I would grow to be exceptional in every way, in accordance with God's will. In the deep embrace of the two godly women, with the lights of the Texas city flickering in the summer night, I was lifted up to Jesus in a prayer of dedication.

Months later we heard the news. Brother Andy had been caught in adultery with his secretary. The Hankses' sixteen-year-old son, who would become a hard-line fundamentalist in the next decade and help plot the takeover of the Southern Baptist Convention, picked up the receiver one night to call a friend and happened to overhear his father in a fairly advanced stage of what sounded like phone sex with his lover. When Andy was confronted by his wife, and then by the church deacons who preceded his dismissal, and then eventually by my father, who, remember, had been best man at his wedding, Andy apologized to his wife and family, the church, the city of Houston, and everyone he'd let down. But he didn't break off the affair and return to the parsonage.

"You know I love Jesus with all my heart, and I love my family," Andy said as my father pressed his old seminary classmate for an answer. "But, Bob, I've got to have a woman who turns me on."

II

Lieber had never heard the phrase "breaking the will of the child." I tried to explain the doctrine—the parent's obligation to bring the child into proper understanding of the child's own sinfulness and rebellion and into a structure of redemptive obedience to the parent. I tried to explain how so many evangelicals believe the child is a creature in active rebellion against all God-ordained authority and the way to remedy that is to destroy the child's natural will and replace it with a newly ordered will, the ultimate goal being a child who almost spontaneously lives into submissive loyalty to God and parents.

Lieber didn't want to hear the theory. Occasionally, yes, he admitted a curiosity about the culture that had formed me, a culture in some ways so alien to his own Jewish community. But mostly he didn't want to hear the theory. He wanted to hear the ways my body and mind had brushed up against whatever the theory was. And so I told him.

Ray Fobb, the youth pastor and youth choir director of my

father's church in Laurel, grimaced and, as if seeking to master some tremendous inner struggle, fastened his hand to my elbow and he led me and two of my friends into the storage room adjacent to the fellowship hall, where the youth choir had been in dress rehearsal for a new musical. I was thirteen or fourteen. Had my friends and I been talking while he'd been working with the sopranos? Mugging instead of taking the lyrics as seriously as we ought to have?

Fobb flipped on the 40-watt bulb. When he told us to drop our drawers, I thought he was kidding; I giggled with nerves. But when he grabbed my left arm and raised his paddle, my smile turned to ice. Since he still had my arm in the grip of his wide hairy fingers, I was first up. I unsnapped my pants and belt buckle, opened the zipper, and then tugged my jeans to the floor.

"Hold your ankles," he said. "Do what I said, hold your ankles."

"I can't hold my ankles. I can't reach that far over."

"Then reach as far as you can."

So I inched my body over and grasped my calves six inches above my ankles, looking over my left shoulder as he adjusted my bare ass and took mental measurement for an efficient hitting angle.

The wooden paddle resembled a baseball bat planed on two sides. When it struck, I went blank with pain, unlike any I had ever felt before. I jumped up, turned around, red-faced and enraged, and said, "I'm going to tell my father about this!"

"Be my guest," he said, and then he spun me around,

torqued me back into position, and delivered two more blows. I wanted to throw up. But I was too filled with shock and shame to do more than tremble.

One night at supper—dad must have been at a deacons' meeting—my mother told me that Ray's brother battled sex addiction; it was why he felt strongly about teenagers making the right decisions. Years later, I heard—from a church secretary, perhaps; maybe from one of the high-school kids—that Ray's wife wore so much concealer for a reason. Was the rumor of wife beating true? I don't know. But it spoke a truth that all of us in civil rights–era Mississippi lived with, knew, and evaded: that violence of one sort begets another. (A few years ago, I googled and found him in the boondocks of the Kansas prairie, where he runs a youth crisis center that provides Bible-based treatment to teens who've succumbed to a sinful life-style. Which sounds like copy for a *Frontline* episode I would not want to see.) In any case, Ray was right about my father. He didn't seem interested at all when I told him about the flogging in the storage room.

It took many years tunneling into the civil rights movement before I could see, alongside the white-on-Black violence, the white-on-white beatings doled out regularly by coaches, principals, choir directors, youth pastors, and anyone else claiming *in loco parentis* rights to other families' children. Increasingly, I see that appreciating the white-on-white traumas of the age—the environing terror that encircled my mental world, my friends' mental worlds—expands rather than diminishes our capacity to attend to white supremacy and its manifold pathologies.

Of course, that Ray Fobb's beating happened at a church was not coincidental. In Mississippi and Alabama, men who hit children are not incidentally Christian—rather, Christianity provides their abuse with an energizing rationale. As illustrated by the seventeenth-century Puritan minister John Robinson, who was a long way from Fobb in time and place, but not in exegetical strategy:

> Say men what they will, or can, the wisdom of God is best; and that saith, that "foolishness is bound up in the heart of a child, which the rod of correction must drive out"; and that "he who spares his rod hurts his son" (Prov. Xxii.15; xiii.24). . . . There is in all children . . . a stubbornness, and stoutness of minde arising from naturall pride, which must, in the first place, be broken and beaten down; that so the foundation of their education being laid in humilitie and tractableness, other vertues may, in their time, be built thereon.

Did Robinson favor a belt? A rope? A hand? Today, practitioners of child breaking like Michael Pearl, coauthor with his wife, Debi, of *Train Up a Child: Child Training for the 21st Century*, are partial to a quarter-inch plumber's supply line, because it fits in a purse or briefcase and can be draped around the parent's neck or left in a visible place in every room and vehicle as a "warning and convenience"—you can buy it in bulk at Home Depot. Pearl, an ordained Baptist pastor, is the founder of No Greater Joy ministries. "The Bible speaks of applying the rod to a child who is incorrigible, disobedient, so I am in good company in agreeing with someone smarter than I am," he once said in an interview, "God." When asked if controversy—specifically media reports that his teachings had been linked to the deaths of at least three children—had increased his book sales, Pearl equivocated.

Corporal chastisement is Pearl's preferred term, rather than *corporal punishment.* "Corporal chastisement is not retributive justice designed to punish the child for the misdeeds," Pearl explained in an interview with Anderson Cooper. "Corporal chastisement is getting the child's attention so that you can admonish him, teach him, instruct him, and guide him in the way he should go." My mother whipped me till I was thirteen or fourteen; like a character in a play, she told me to go outside and find my own switch, and then she'd hit me good with it. And countless coaches and school principals hit me, in the classroom or more likely in their cinder-block offices, during or after practices.

How did it come about that violent men with no training as teachers or preachers or counselors made their way into positions of power over junior-high and high-school students? Some played football for Bear Bryant, which in Alabama in the 1970s was as close to apostolic succession as you could get. Others arrived with dynamic testimonies—which is to say, they'd recently overcome addiction. Some ended up in youth ministries, it can only be surmised, because they had no other career options.

Sometime during my years with Lieber, I began reading Donald Capps, a professor of pastoral theology at Princeton

Theological Seminary, on the nexus between spirituality and the physical abuse of children. American practices of corporal punishment, Capps argues in his book *The Child's Song: The Religious Abuse of Children*, are inextricable from Christian theologies of atonement and from an apocalyptic worldview that turns on images of punishment and tropes of submission. But perhaps more helpful to me, as I strove to understand the roots of my mental illness, Capps argues that corporal punishment, especially when meted out by a religious authority figure, can create pathologies of shame and anxiety.

Beatings teach children that their bodies are inherently corrupt, that their bodies require only discipline and never admit joy. Beatings ask children to bear in their bodies punishments that hint at the apocalyptic punishment they might one day face. And beatings that surround children's bodies with the language of disobedience—disobedience whose prepositional phrase easily slips from earthly father to heavenly father—ask children to respond to God with an awakened fear, the same fear children might feel at the sight of a belt or a paddle or a switch or even a hand.

Is it any wonder, I thought, reading Capps, that I came to expect that desire would issue in punishment? That I came to expect that if I acted spontaneously or said something risky or pretended I believed I was free, I'd be met with pain?

III

Another violence I handed off to Lieber was the violence I meted out. I went to elementary school, in the last years of the segregated system, with a boy named Jacob Krawley. He came to school unwashed and smelling of wild onions; he looked like a mangy dog, and we treated him like one. There were burns down his neck and shoulder; there were often black eyes and bruises. He was about two years older than us, because he'd been held back at least twice. Jacob's father was a Klansman with a law degree from a mail-order law school in Jackson; he was a drunk and, of course, a child beater. On the playground, we'd circle around Jacob, and chant, "Bo bo bee, O Krawley, that is who we want to see." It was a ritual. Some days he just recoiled, like that scared dog, but sometimes, with a pocket knife or pipe, he fought back. Mostly he just cowered and began crying. Once we had broken him, we went off and played kickball.

I might see him in town. (Did he attend a white working-class church in the county? I doubt his father worshipped anywhere.) I might seem him in the hall at school and feel tender toward him and ashamed of my behavior. I wanted him to be loved. I also couldn't stand the sight of him. Something about his presence felt as though it needed to be removed. How does one analyze that apart from the closed society? He was dead at seventeen.

IV

Haltingly at first, I talked with Lieber about adolescent sex—which is to say, I talked about masturbation. I was late in discovering I could give myself a wet dream, I told him. One night after basketball practice, I was soaking my body in a hot bath when almost effortlessly a load of semen blew up through the suds with the force of a whale's spout. Coming at last into this new knowledge helped explain things I'd heard kids my age say, things that had not computed at the time. "Find a damn stall," a boy said in the locker room. I didn't know why. I felt good about the erection; the erection made me feel as though I was going places. But a middle-of-the-night surprise was the best I could hope for. Now that the truth was revealed, I brought a convert's zeal to my devotions to Rosie.

Like other evangelicals, I assumed that God was extremely concerned about my genitalia. God had first dibs on it all. He'd designed the whole package, shaft and sacks and valves. To be used in the operations of holy matrimony and piddle and nothing else. Premarital and extramarital sex got him angry for reasons I could never remember. And, let's be honest, if you read the Gospels closely—as closely as an evangelical boy in the throes of puberty—you won't find Jesus worrying much about sex, though you might linger over the sight of a woman rubbing oil into his feet and his approving observation that she "has not stopped kissing" him.

Did the Messiah not pleasure himself? You'd think he would have, if only once, since the ancient creeds say that in Jesus God became fully divine and fully human, with anus, testicles, and all. I'm simply following the lead of an early church father named Tertullian, who first called attention to Jesus's descending colon. The Word made flesh is messy business.

Upon making the bald man cry, I flushed down the results with a prayer for forgiveness and the bold resolve to never again succumb to the right hand of sin. I recited the verse about thinking only on things that are right and pure, and the one about offering up your body as unto the Lord. And then I treated the tube sock to more of my racket and swoon. Psalm 51 became my go-to confession: "Have mercy upon me, O God, according to thy lovingkindness: according unto the multitude of thy tender mercies blot out my transgressions. Wash me thoroughly from mine iniquity, and cleanse me from my sin. For I acknowledge my transgressions: and my sin is ever before me." By the time I graduated from high school, I knew all nineteen verses by rote.

One summer at camp, I'd abstained for six weeks, far and away a personal best. It was a fleeting triumph. As I rode home with my parents at summer's end, the heat of the day and purr of the lonesome highway lazed onto my bucket seat and produced a sudden thunderous erection. I managed to reign it in slightly, but as soon as we approached the outskirts of Meridian, I asked my father for a bathroom stop. The welcoming sight of a Bonanza Steakhouse appeared in the near distance. He looked at his watch and said, "Why don't we get lunch too?"

Inside the restaurant I slipped into the bathroom without incident. It took so little effort to fail that day. I'd hardly done more than position my penis over the urinal as if to pee, when joy shot from my loins so fast I could hear sirens. After all the good work of the summer, I was once again captive to the carousel of my wanking life.

I felt blindsided by desire and by the possibilities of what I could do even during a quick bathroom break once liberated from the security of summer camp. Whatever pleasure there had been in the orgasm gave way to a panic about how much my body and, of course, my mind remained in disobedience, in revolt, in a state of inescapable failure to live up to Christlikeness. As a camp counselor, I'd led many an unsaved child, lost and hell-bound in some lonesome pine outposts, to new life in Christ. And I suppose I had formed an unconscious notion that I had prevailed over the flesh, that a few weeks of digging deep in the Word and reveling in the admiration of people who esteemed me as a young Christian man had transformed me. The Bible kept insisting, after all, that I would be transformed—that I would be remade into the mind of Christ, that I would die to the rudiments of the world, that I would put on the new man. In a flash that was gone, replaced by the knowledge of myself as just an ordinary horny sixteen-year-old who'd ejaculated into a urinal in a steakhouse outside of Meridian, Mississippi.

What I began to surmise in Lieber's company was this: My later breakdown was in part a response—a fair response—to the visceral sense I had that afternoon, and so often in my adolescence, of my body's disgrace, of my body as ground zero in a warfare between holiness, on the one hand, and the world, the flesh, and the devil, on the other. The breakdown was an eruption of a body that was in a state of war with no promise of cease-fire on the horizon.

In Lieber's office, I was able to demythologize all this—to drain the language of spiritual warfare of its force. I was able to see—to begin to see—that desire is not a whore. Desire is sorrow in my heart. Desire is awakening to the lack that "comes howling down Elysian Fields like a mistral." Desire is a delirium, a frenzy . . . the energy of life. Desire is the human situation.

V

Violence, sex, punishment, hunger. Of course, I also spoke with Lieber about my father, the strapping young preacher, Bible quivered in one arm and the other arm jabbing the enceinte air of the pulpit. Resolute in his seersucker suit, turning to the congregation, and sometimes scouting me out in the balcony, more summons than scorn—*I am talking about eternity down here, son (I know, Dad, it's feeling like eternity up here)*—he gave my struggle its punch.

Early in his ministry, though not so early that I can't recall them now, he preached sermons straight from the church's catalog of horrors. Sometimes from a literal catalog. He received a weekly supply of preacher's aids, sermons in outline, and sermons on tape and, like most of his colleagues, happily availed himself of their inspiration. I doubt that the hell sermon preached by the evangelist to a klatch of pubescent boys in James Joyce's *Portrait of the Artist as a Young Man* numbered among the SBC's resources, and I never saw the novel in my father's library. All the more striking that my father's efforts followed the same game plan, punctuated with a homiletical ferocity worthy of Billy Sunday or R. G. Lee, whose popular sermon "Payday Someday"—which Lee claimed to have preached a thousand times and to more than a million souls—my father could recite from memory.

"I would rather not speak to you of hell," he would say. "I do not like the idea. It gives me no pleasure to expound upon the torment that awaits every sinner's appointment with death. But my feelings concerning a fact will not alter a fact. . . . And hell is a fact—of that I must remind you. Hell is an eternal place of everlasting separation from God. Hell is an eternal place of everlasting anguish." My father spoke of time running out, and of hell's being our destiny if we did not make the decision for Jesus while we still could. "Jesus Christ stands today as your friend and Saviour [*sic*]," he preached. "One day He will have to say: 'I could have saved you and redeemed you if you had come sooner, but now I can only judge you.'" What would we then "give if only [we] could come from the pits of hell and have a little more time?"

When my father concluded the sermon, the invitational hymn began to play. At that moment each week, I could feel my soul melt into the perfect sadness of submission. "All to Jesus I surrender, all to him I freely give; I will ever love and trust him, in his presence daily live. I surrender all, I surrender all."

Surrender was one of first words I learned as a child. I sat beside my mother while she played the late-nineteenth-century hymn on the piano. I hummed it as I moved toy soldiers (Rebels and Yankees) across the green jute rug of my bedroom. I felt its weight.

Fast-forward a half century, to a January afternoon in Bologna. Taking advantage of a winter special at Lufthansa, I'd cashed in a modest number of points for a ten-day post-Christmas vacation with K and our two older children. At the National Gallery, I stand transfixed before Maestro dell'Avicenna's *Paradiso e Inferno* (1435). There, set against the grisliest hellscape ever, Satan feasts on the bodies of the damned—he shoves some into his dark throat as though they were stiff lizards, while others exit his nasty red ass into the pit. Jesus H and Josef K! Who even knew that was his thing? I wanted to write a note to Lieber with a late revision to our work: "I still think I'm going to wake up in hell, just like the people in the painting." Something like that.

Instead, I fired off an email to a colleague at Yale, a scholar of such things, whose Pasolini-like fascination with Christianity's most bizarre habits has always comforted me. The year I'd worked as his teaching assistant was a rare glint of clear sky in the purgatorial years of grad school—and in the decades since, we'd met up for drinks in New York, cohosted academic soirees, and become friends. Fond of hovering nuns, floating friars, and other levitating early moderns, Carlos could always be counted on to find humor in weird religious behavior. About the afterlife, on this day in Bologna, he didn't let me down.

"That depiction of hell is horrific," he wrote back, "very typical of the late Middle Ages. A splendid thing, no? Hell changed in the modern age. Teresa of Ávila was one of the first to describe the modern hell: no physical torture by demons, just a tiny cramped niche where one would rue one's sins for eternity. Psychological torture in a cramped space. I might rather prefer the shitting beast."

Thanks for that. I stare at my cell phone, as I now stand in the Piazza Maggiore in a light snow, and say a silent prayer for oblivion.

VI

And, of course, my mother. My mother I sometimes think might have become the American Karl Barth or the southern

Marilynne Robinson, had she not been saddled with the patriarchy of the age. She was valedictorian of her senior class in high school as well as Most Friendly, Most Versatile, and Miss Central High of 1953. She finished three and a half years at the Presbyterian women's college with straight As and the praise of her teachers. My mother so impressed a visiting theologian from Edinburgh that he read sections of her essay in his public lectures on Reformed theology.

These maternal could-have-beens demand a reckoning with the tragic costs of Christian patriarchy. I thought of my mother recently when reading an interview with Rebecca Traister:

> We can't imagine what the world would've looked like if this systemic behavior hadn't been in place. We don't have the buildings that were built by women or the food that was cooked by women or the comedy that was written and performed by women or the art that would've been made by women or the books that would've been written by women or the political narrative that would've been told by women or the candidates and politicians and political leadership that should've been female.

Traister had in mind both Christian patriarchy and the last lines of *Middlemarch*.

So hemmed in. I hear the same laments from friends. "My mother would have been governor if she'd been a man," says one. "My mother would have gone to medical school if she'd been a man," says another, "or if she'd been born forty years later."

Did my mother know of those few models of female power, the radical white women of the South? Of Lillian Smith, with her rescindent insights into the logics of segregation and her summer camp that aimed to "work with girls who will someday be the women leaders of the South"? Of Lucy Randolph Mason? Of Jessie Daniel Ames?*

Of course, she did know of Eudora Welty, whose English Tudor cottage sat on the hillock across the street from Belhaven College. Of course, she did know of Elizabeth Spencer, a recent alumna who published her first two novels shortly before my mother entered Belhaven. And of course, she did know Ella Ransom, sister of the Agrarian poet; my mother took every English literature course Miss Ransom taught at Belhaven and credits her, if not with seducing her into professional life, at least with bolstering what would be her lifelong love of literature.

Over and over, I unspooled for Lieber how, in the fall of 1956, submitting to patriarchy meant my mother's leaving college one semester short of graduating to move with her new preacher husband to Mobile, Alabama, into a house of cramped,

---

* Still, none of these regrets can change the fact that, in the South of my childhood, white women, with a few glorious exceptions, exercised individual agency with a villainy every way equal to that of white males, standing alongside their men in as guardians of the closed society. And it was usually the mother who taught her children "the lesson," that although you may one day find yourself playing kickball at the park with a Negro child, you can never be friends, etc., etc. And this is why it does not occur to me to ask, did my mother know of Ida B. Wells-Barnett? It is why Margaret Walker, teaching at Jackson College, living in those years with her husband and four children in Jackson, writing poetry and fiction, could not then have shaped my mother's imagination.

musty rooms, overheated in winter, hot as a steam bath in summer, whose darkly painted walls and velvet drapes produced a chilling effect without lowering the temperature. It meant putting down the Dutch Calvinist tomes on doctrine and the Agrarian poets, books she loved in equal measure, to serve the needs of her husband's working-class parishioners—to prepare casseroles and bake pies when a visiting preacher came to town; to sit with the women of the congregation as they mourned, told her their secrets, asked her for prayers, worried about their children, and leaned on her conversation to give shape to their days—and to surrender too to her own panics and depression, which lasted too long.

Summer in Laurel

Everything I'd read about finishing analysis made me want to avoid the subject for as long as I could. Lieber's estimation that a complete analysis would take about three years seemed increasingly implausible. When year two came to an end, the question of termination arose. I mustered up the courage to say, in so many words, I'd prefer not. I'd been away both summers on research, I said, losing months there, and wanted the full experience. Lieber, of course, was interested in hearing more about my concerns, my fears about termination.

When he asked how long beyond three years I imagined analysis might last, I said, "A year and change." I feared that he would disapprove and stick to the script, but he seemed unfazed by my request.

We talked again about fees. I'd made money on a book prize, how much I didn't say. We agreed that it would be fair to raise the rate to ten dollars an hour. I didn't tell him, although maybe he knew, that the phrase "and change" allowed me to imagine that we could continue treatment indefinitely. It was my best effort to say, quoting an old *New Yorker* cartoon, "How about never? Never works for me."

In *Psychoanalysis: A Very Brief Introduction*, Daniel Pick writes: "Where analyses in Freud's early days might be over in weeks or months, particular subcultures of analysis developed (worryingly to my mind) to the point where they can endure routinely for many years, with insufficient questioning of the rationale: an assumed way of life, rather than a mode of treatment." Pick thinks this insufficient questioning can be exploitative, detrimental to the work, even a *folie à deux*, a shared madness between doctor and patient. In some cases, patients may end up "nursing" their analysts into their twilight years, so agonizing is the decision to retire. Such patients are happy to dither and stall into the eschaton. And can you blame them? *Termination?* A less tender benediction is hard to imagine.

Another year passed. Another eleven months of near daily commutes into the killing fields of east Baltimore. I'd begun thinking of myself as the Bob Wiley of the Johns Hopkins psych department. Remember Wiley, the analysand who wouldn't leave in *What About Bob?* Who stalked Dr. Leo Marvin to his New Hampshire summer vacation home and spoiled his August hiatus? "I need. I need." In Bill Murray's gloriously clingy lollygag, Wiley is, in Dr. Marvin's estimation, a "multiphobic personality characterized by acute separation anxiety and extreme need for family connections." The tables are turned, of course, as you know if you've seen the film (and if you haven't, you should).

If you reach this place, as I did, you need not be discouraged. There can be comfort in knowing that you don't want analysis to end, that although you've become a force of a person in ways inconceivable a decade ago, the cause of your illness is both deeper and more identifiable than the talking cure could ever fathom. But that was cold comfort at the time, as I circled endlessly around the same old shit with no end in sight. I'd kind of always known I would never pass through the termination stage like the suburban analysands of the 1950s and 1960s and that my cure lay somehow in reckoning with this fact. With the fact you're never finished with the work—that there are epiphanies and resolutions but there is no closure—and that the only termination in the otherwise interminable blah, blah, blah of it all is death.

———

Still, my last session with Lieber did come, on a bright Thursday in November of—well, let me put it this way: a bright Thursday well more than three years after our first consultation. What, in the end, were the fruits of the hundreds of hours in his office? Perhaps the best way to answer that is to tell you of a summer that could not, would not have happened, were it not for our work—an earlier summer, in the middle of analysis, when I went back to Laurel, muscling my way into a tangle of memory and fear.

I was there to interview both the saints and the villains of the civil rights movement, to unearth the theological commitments that animated their politics and activism. One afternoon, I drove to Jackson and luxuriated in the magnificence of Lemuria Books. From the history section, a glossy white cover called out to me: *We Are Not Afraid*. It was the very book on Schwerner, Chaney, and Goodman that I'd not only been unable to review, but even to read all those years ago in Charlottesville. I bought it, took it back to the house K and I were renting, and read it by the weekend.

Another thing about that summer in Laurel? Sex. I remember a week when I'd been on the road for several days and, seeking only K for the weekend, arranged for our children to stay with friends who owned a catfish farm near Hot Coffee.

I had been so inhibited, simultaneously both bound to and disconnected from my own desire. In Lieber's care, I gradually became freer, and somehow more unified. I could talk about sexual fantasies. I could imagine, aloud, fucking the woman with whom I'd sat in French class so many years ago (every morning, she, in a skin-tight miniskirt, often leopard print; class was held, I kid not, in Cocke Hall). I could imagine fucking my mother's next-door neighbor on the kitchen table where she so frequently served my mother tea, both of us, in my mind's eye, still mostly fully clothed. I could offer Lieber the heaviness of all that adolescent masturbation and the shame it carried. I could talk about these things and see, in time, that all this talk wouldn't kill me, might not damn me, might unfetter me even. I could talk about Sheila, whom I still saw occasionally at conferences.

I told Lieber the story of a trip I'd made south a few months before beginning analysis—a road trip through Georgia and Mississippi that would culminate at one such conference in New Orleans. In Laurel, where I'd stopped to see old friends on Sunday, I attended the church my father had pastored, the aqua wall-to-wall carpeting unchanged, the massive pipe organ. During the service, I became totally overwhelmed, somehow at once a man in his thirties and a boy of twelve, trying to appear attentive to the sermon yet unable to take my eyes off legs and thighs and hemlines. The gospel was proclaimed, people sang their Baptist hymns, everyone in their finest clothes, and I had a lunch date at the country club following the service. But I felt as if I was choking, my heart pounded louder than the organ, and I was sure that everyone was looking at me, scrutinizing my fidgets and my sweats. I was again a child, feeling that I was supposed to be growing in Christ, but the only thing growing was my sad pubescent dick.

And the only way I could moderate the symptoms that Sunday, the only way I could halfway begin to breathe, was to go to a sex fantasy, to remind myself that in a few hours I'd drive to New Orleans and have ~~sex~~ dinner with Sheila (who was now married, though not to Miles). The only way I could get through the service was imagining that later that afternoon I would see Sheila and go down on her. I would eat her till she screamed for me to stop. I was on the road to New Orleans by three o'clock, and with every mile I felt better. I did see Sheila that night—but that's all, only saw.

My ability to talk about sex with Lieber allowed me, in time, to bring my own desire out of the forbidden margins of my life and place it nearer the center—to bring it, that is, out of the realm of fearful fantasy. And, of course, once I began to see the ways the cycle of fantasy and shame and hiddenness had served to keep me from integrating the erotic into my life, I awakened to something I'd not allowed myself to appreciate before—to the pleasures of sex with K. To the erotic presence of K and the "effloresce of new insights," as Peter Gay rendered the blossoming of carnal knowledge in nineteenth-century Europe in his marvelous book *The Education of the Senses*. To the ways that sex with K could be at the center of my actual life, could exhilarate my real life, and the knowledge that such sex needn't follow any script or be vanilla or pristine.

And K herself was looking for more than the forthright missionary communions that had attended our marriage until now and produced our three children. She too, it turned out, wanted to tease me, say, as we dressed for a birthday dinner with friends, with hints and phrases of what awaited us at midnight when we returned, and she too wanted me to feel her up in the back hallway of that restaurant by the pay phone, and she too wanted me to look at her across the table and think, with a kind of shock.

One weekend, shortly after that Laurel summer, we assignated as if we were near strangers having an illicit fling. It was New Orleans again, a small hotel on Esplanade. I'd been on a research trip in Neshoba County, and I would meet K at the hotel, coming from a conference in Jackson.

Outside of the city, I run into a thunderstorm. The spring day has been warm, and I am not surprised to see the dark clouds moving over the Pontchartrain and a blue-gray sky descending on the city. The water hits my windshield like the sudden spray of a car wash, and I lean forward in my seat, straining to keep the highway in view. The wipers turn at max speed, but the force of the rain is too strong for them to clear the glass, so I slow the car down and turn off the radio; other cars have pulled off the road and are waiting the storm out, their emergency lights flashing. In the northbound lane a slow parade of automobiles moves away from the city.

I know that my wife waits in a room on Esplanade. She might be napping now or reading a novel on the sofa. I hope she got the flowers I had sent to the room.

The rented SUV feels good and safe; the heavy wheels of the vehicle hold strong to the asphalt and cut through the flooding highway with ease. I pass the road signs for Gentilly, a neighborhood where my relatives had lived years ago and where, as a child visiting from Mississippi, I had felt happy and free. The exit to Elysian Fields appears soon enough, and I drive down the ramp into the low streets near Armstrong Park.

On so many driving trips through the South, I had fantasized about a dream girl joining me at the end of the day, the two of us locking the door of our motel room and stripping away every inhibition and every remnant of decency until all that is left is the sweet odor of flesh.

And yet now, as I pull the SUV onto Esplanade and see the hotel in the distance, my pulse quickens in anticipation of my wife's flesh and of her sex. Once out of the car, I ascend the staircase as fast I can, the unexpected rush of temptation bringing me to the one place I never really expected to find it. My wife is sitting in a chair wearing a green silk dress and holding a flower in her hand—under the fifteen-foot ceilings, the velvet wallpaper buckled by humidity, the oriental rugs with cigarette burns—and she is smiling wickedly. Oh, sweet temptation, thy name is K. Look at me, I have survived the long ordeal of my sorry old self, and here I am as horny as a deck hand, a man reborn. It will take a lifetime to know her the way I should, so let's get things started right now.

———

The years of analysis notwithstanding, that summer in Laurel I found myself beset by a dream I'd had often enough during the previous decade, about a town named Sad Cotton, a fifteen-minute drive between Downinout and Ohnwee if you're doing the speed limit on Downcast Highway #9. It's nowhere near a hospital with a psychiatric unit or a decent psychotherapist other than the kind you're likely to find at the Logos Counseling Center, the last resort for the churchgoing middle class.

Sad Cotton is no howling backwater like Tarbottom or Humpjaw. I'd be surprised if you couldn't find a renovated barn or farmhouse-turned-hub for local arts. In the dream, I've moved here with my wife and two sons; we found a bungalow; we have Jefferson cups for juleps. I'd chosen to return and make a life here after finishing law school in Virginia. My classmates took jobs in big cities with posh firms and nicknamed me Atticus when they learned of my plans. But my decision had little to do with a concern for minorities, even though I considered myself a liberal and choked up whenever *To Kill a Mockingbird* aired on TV.

I appreciate the town's proximity to the coast; I want a lifetime of warm autumn suns. I plan to take simple cases, leave the office early and enjoy the evenings with my wife and children. So I take a job with a firm on a court square in a county seat, renovate the bungalow in the historic district, and join one of the Protestant churches downtown.

However, a problem presents itself early on that cannot be resolved, and the problem gets worse as the weeks turn into months. I have begun feeling afraid, not of anyone in particular, but of the thought of myself in this town, a town that very soon felt strange and menacing, and afraid of myself as a self, a mystery without meaning, a body without sense and order. The dream turns to unabated dread, made worse by the reality of the pleasures my dream self had imagined: the October sun, the wind in the pine trees, and the solitude of the early mornings. I had never fired a gun since childhood, but, with my sons wrapped around my wife in sleep, I retrieve a hunting rifle borrowed from a colleague out of the trunk of my car, slip into a shed beyond the abundant garden, open my mouth around the barrel, and pull the trigger.

In Laurel, I dreamed this dream, which I'd imagined I was done with, three nights, four nights, five. I considered how lucky I was that the fates moved me to Baltimore, to a city. Were I in fact a small-town attorney in small-town Mississippi, my mental-health options wouldn't have been much different than if I'd lived in sixteenth-century Europe. There I would have seen a priest for a blessing, been instructed to make a pilgrimage to a shrine, and finally had to concede that God "can stoke [us] with madness," can "plague us by his own design, creatures, sun, moon, and stars, which he useth, 'as a husbandman doth . . . a hatchet,' whomever he will." Over terrors aroused by sin, my condition would have gone by the name of religious madness, which is precisely how I received its offerings during my first breakdown.

Are there white Southerners without suicide in their lineage? Perhaps I'd just read too much Percy, too much Kate Chopin and W. J. Cash, seen too many productions of *Crimes of the Heart*, but I'd awake from that dream and know I came by the expectation of suicide genealogically, which is to say I woke and thought of my great-grandfather. How unspeakably shattering to find yourself mentally ill in the 1930s in Lumberton, Mississippi. His health deteriorated so rapidly toward the end of his life that his daughter left the music conservatory in Atlanta, despite being only a few months shy of her diploma, to look after her father. Even so, when, on a spring morning in 1931, she heard a single gunshot from the front porch of her family home, the sight of her father's body left her grief-stricken but not completely surprised.

He was buried in a field of wildflowers next to his wife and two infant sons. In turn, my grandmother battled deep blues herself for a lifetime. And Nana never talked about her father's suicide, neither with her own children nor her closest friends—and definitely not with her inquisitive grandson. (Indeed, I know little not only about his death, but about his life; I can't tell you even now why a melancholy Swede, a chemist by training, left his home on the coast of Ronneby and settled in Lamar County, Mississippi, to run a drugstore. The closest I got to answers was the wife of a second cousin in Hattiesburg, who'd studied the history of the Ollsons of Ronneby; but by the time I called on the cousin and his wife, they were separated and not speaking to each other, nor to me.) For the young woman with Old World loves, peace of mind lay in a sure refuge; she craved a life so ordered as to shield her from complication. Late at night, I could hear her footsteps in the halls of her home. She was making sure the bedroom doors were open. She did not want to talk about her past, and she didn't want to hear about anyone else's either. Her sighs were as constant as the stirring of ceiling fans.

———

The book I was writing—well, I didn't know precisely what the book was when I arrived in Laurel, but it began to take shape there on the hot afternoons in the rented house or in the marbled reading room of the museum a short walk away, in lists of characters and scenes and in words, phrases, and memories meticulously recorded on a yellow legal pad. Looking directly at the closed society had once terrified me, but I found now that I loved the feeling of being an armchair agitator. And I loved even more the feeling of unearthing, and then telling, people's secrets. Sifting through boxes of letters in former neighbors' attics; hounding Sam Bowers for an interview; tracking down Jane Stembridge, who was now living in rural Illinois; pressing the manuscripts of sermons from the 1960s for their strained segregationist logic—all of this, I felt, put me somehow in the company of Victoria Gray Adams and Walker Percy. It was Percy who, in a remarkable essay, "Mississippi: A Fallen Paradise," laid his adopted state if not on the analyst's couch, then on the fainting couch, hoping to understand its neurotic loyalty to a humiliating past.

My work was an imaginative crossing over to them—my particular way of contributing to the politics of liberation— and an unexpected realization of writerly fantasies I'd harbored since high school. I discovered in myself an assertiveness I didn't know I could summon, hounding people until I got what I wanted—an interview, a peek in their filing cabinet, the answer to the question they most wanted to avoid. I often thought of Joan Didion's insight that her "only advantage as a reporter is that I am so physically small, so temperamentally unobtrusive, and so neurotically inarticulate that people tend to forget that my presence runs counter to their interests." It was a thrill to feel that I'd been let into her secret.

Anxiety symptoms continued to surface at times, as I expected. Sometimes, sitting in a library, I'd feel the old inability to breathe, I'd feel the panicked sense of being stared at by all the other researchers, or I'd feel my legs go numb. What was new was that I could barrel through. I had a repertoire with which to organize the symptoms like characters in developing narrative, elements of story; the symptoms less and less organized me. What was new was also the company in which I now found welcome and hope. I was beginning to think of Fannie Lou Hamer and Chaney, Goodman, and Schwerner as civil rights saints, never mind their not being judged so by any standard ecclesiastical metric.

My parents had long since left Laurel for Atlanta, where my father took the reins of the Second-Ponce de Leon Baptist Church, a white-brick, fifteen-acre fortress on Peachtree Street and gateway to the city's affluent Buckhead district; he'd made the big leagues at last. Still, the book was in some way my parents' story—the story of ordinary women and men caught in the whirlwind of the movement, the ways the faith sometimes buckled under and sometimes pushed against the etiquette, intimacies, and violence of Jim Crow. That summer, I spoke with my parents relentlessly about our tenure at ground zero of "the worldwide mongrelization movement" and my own coming of age in Jim Crow's brutal pigmentocracy—in email exchanges and phone calls and throughout their lengthy visit with us in Laurel (their two darling grandsons being the main attraction).

At first the conversations were framed with reference to my historical research, but they fanned out and fanned down. I wanted to know why so many people in my family were depressed. How did a melancholy Swede end up in a settlement etched into an empty canvas of virgin pines and pecan trees sixty miles from the Gulf? And why, then, did he shoot himself? Didn't his leather-bound book of patents include recipes for the tonics of the era that promised to cure melancholics and enliven sluggish days? (Had I been a regular at the Pearl River Pharmacy, I might have been partial to the Brompton cocktail. I like to think that Martin added a mint leaf to this epic swirl of cocaine, morphine, Thorazine, and gin. Holy shit!)

I never raised my voice to my parents—never said to my mother, "You shouldn't have told me those things! Look at what toxic fruit those fables you told me bore!" Instead, I mostly asked questions. "Why were you so upset when you learned that your father had been married before he married your mother?" "What were you afraid of?" "Did you never stop to think about—oh, I don't know—Jim Crow racism? How could y'all keep whirling and doing and praying and giving and never stop and think that maybe it would be okay to let the yard man use our bathroom? More than okay, even. Hospitality to strangers, angels unaware—ring a bell?" I asked even though my historical study and my loyalty to the doctrine of original sin both punctured my pretense that I would have behaved any differently.

Later, I saw that I had of course learned this from Lieber—how to ask such questions, how to bring into the light by questions, how to unshackle just a little truth by posing the questions again and again.

My father still wrote me letters that summer: "Remember that no one ever loved a son like Bob and Myra Marsh loved their son. Know that no one ever sacrificed like we did. All we ever wanted was for you to live up to your potential. But go ahead and ask your questions, son."

More remarkable than acknowledging my asking, my parents began to answer—unlike most white Southerners I've queried over the years. Sometimes in conversation on the phone or around the table, sometimes in parcels of old letters, journals, sermons, and family esoterica that arrived at my homes in Baltimore and Charlottesville in the months and years to come, my parents found their own courage, and they shared that courage with me.

——

Nana still lived in Jackson that summer my family and I returned to Laurel. It delighted me to see my two little boys scrambling around her back yard like puppies and eating the pecan pies and sugar cookies whose sweetness is one of my earliest memories. My grandmother didn't much leave the house those days, but my addled spirit carried her with me on all my summer interviews, shucking corn with Will Campbell on the porch of his writing shack, or eating oyster stew with Bob Zellner at Casamento's, or slogging into early morning heat to meet John Perkins for breakfast at Shoney's Big Boy.

It wasn't enough for these righteous giants to give me their time and tell me their tales; I wanted something more from them. I wanted them to make me fearless—the way I felt when we were together. Fearless and joyful and sure of where I stood. And I wanted them to explain things to me. My grandmother, for example. Who'd recently instructed me as I was heading out for a day of interviews to tell that man so-and-so that, no matter what he thinks about all God's children, white people are the superior race. I wanted them to say something that would unburden me of the words she tucked into my pocket every time I paid a call.

But my interlocutors, who'd ordered their lives around their commitment to justice and their insistence on seeing humanity where other people did not, weren't going to relieve me of my guilt; they would neither implicate me in nor absolve me of the taint of familial sin. Instead, Will Campbell, who liked to put tomatoes and peppers in his cornbread, gave me a bushel of cornmeal to take to Nana that weekend. Instead, Zellner insisted I carry a barrel of oysters over to Nana that very day. Instead, Perkins—to whom I'd wailed about my grandmother at no less than four separate crack-of-dawn breakfasts that summer—said, "You know, Charles, how I love blueberries." And it was true. He loved them on ice cream. He loved them in a cup of milk with a little dash of vanilla and sugar. "I have some blueberries in the freezer—when we get back to the house, I'm going to fix up a bag for you to give to your grandmother."

Please indulge me, dear reader, this grace note. Many years after I gave those berries to Nana, my parents sat down with John Perkins at a cafeteria for a meat-and-three and passed most of the afternoon together, talking about their ministries, and scripture, and their children, and their mothers. They held hands and prayed together and talked about aging and death. They wrapped up that meal with cobbler—whether it was blueberry I do not know.

Years of Wonder
and Longing

Analysis met me where Freud said it would: in work and love. But what pushed me onto the couch in the first place was not, how shall I say, the magical thinking that analysis would transport me to a place where the orgasm never ends, nor was it even the more proximate hope of becoming a better marital lover. It was not a desire to read or write books about Freedom Summer. It was not, principally, to fathom the recurrence of certain words and images in my notebooks, words and images that evoked the night terrors, in the claustrophobic bedroom with the burglar bars and brown shag carpet. What brought me to analysis was, instead, first and last, a wish to father my children well.

———

I sit in my study trying to read Alasdair MacIntyre's *After Virtue* for a faculty seminar at the college. My thoughts drift toward the coming winter. As cold winds rattle against the window: *How harsh will it be? Will I remain a reliable presence? Why am I living this life?* Only last week the air was unseasonably warm. I went jogging in shorts; Henry and I played baseball in a field near our house. But tonight, I placed extra blankets on the children's beds. I turned the heat up a few degrees, and I stuffed the drafty window seals with rolled-up towels. My ears are still sore from this afternoon's run at the Hopkins track.

I try to turn my mind to warmer places, the towns I lived in as a child. I call an old friend from Alabama who lives now in Orlando and ask him to describe the weather. I open my calendar and begin making plans for a midwinter trek to see him, but discover that every weekend through May is filled with kids' sports and work.

I am drawn to a Faulkner novel that has sat on my desk for months, *The Wild Palms*, with a cover of seagulls sparked against a honeyed summer sky. I want to disappear into a honeyed sun and a brown-shingled bungalow at the end of a white shell driveway. Alas, the two main characters will have none of it. From what I can make out—and don't hold me to it—they move from Biloxi to Chicago, inhabit an abandoned house, and are sent, by the midwestern winter, into a death spiral. But then there's a flood—the Great Mississippi Flood of the year I've at the moment forgotten—the bright glitter of water; the stagnant, opaque, flat silent sheets of water; and I'm lost.

I turn to the Bible, letting it fall open the way the eighteenth-century enthusiasts did. There's nothing random in the results. The suffering servant shared all our afflictions, took on our infirmities, shared our sorrows. By his stripes we are healed. *His stripes. Not mine.*

The snow is gathering force outside. I'm reading Isaiah, and

I'm thinking of Jesus, and how he suffered, but never a long winter's night on the East Coast.

———

I read Francis de Sales's *Spiritual Directory*, in a worn leather edition I found at the library spring sale. It sports a thin gold ribbon, which leads me to a page stripped to a single thought: if we awaken during the night, from the view of darkness around us, we should turn to a consideration of the darkness in our soul and in all sinners and offer a prayer. Oh, my dear fairly forgotten Francis de Sales, edged out of the spotlight by the vigilantes of self-loathing, I wish to consider my darkness in love.

———

The blue sky spreads like a warm quilt over the day. I open the window to the hissing of insects. Early summer is as soft as a child's hand.

My daughter narrates her world as she experiences it. "I'm stuck," she says. When I help her through, she adds, "I'm not afraid of guinea pigs." You would need to know she's referring to her brothers' new pets, given weeks earlier as Easter presents. One is named Rifle and the other Musket. Not knowing this might be even better. God broke death's door on a criminal's cross; now go and collect your fears in singsong defiance of spiders, bats, and—why not?—guinea pigs. Be not afraid.

———

I would not raise a child on a diet of battering words. I would not use the Bible as a bludgeon. I wanted my children to learn a language of peace, to sleep at night in the knowledge that they are loved, with a fierce and indefatigable "Yes." The only risk I wanted to take in my love was one that threatened excess. I wanted them to know that that their aspirations are noble, and they can trust their bodies, their minds, their desires. I could not immunize them from life's uncertainties, but I could relieve them of the weight of evangelical anxiety.

———

Will and I attend the Sunday evening Eucharist at the Episcopal church near the university. The windows in the chapel are raised to a warm breeze. The priest speaks briefly of the liturgy in the ancient church. He says that the reading of scripture often lasted over an hour, and that listening to the words created a different kind of intelligence, an intelligence of the heart.

After the service ends, Will and I walk home together through the summer night with not much more than the roar of cicadas to occupy our minds.

———

Sometimes I awaken in the morning to a felt awareness of myself as God's beloved, but I am hardheaded and much too receptive to every contrary emotion. I need that awareness written in letters tall enough to read with confidence, although it rarely happens that way. When the awareness comes, it's usually a simple blessing easily missed: the stillness of a Pennsylvania forest, where I have rented a cabin for the night, having delivered my two sons to summer camp; the sky a sea of blue; birdsong; echoes of quiet stirrings on porches and campsites; the smell of wood smoke from an old man's fire; my glad and somber heart on this July evening.

Two weeks later, when I return to pick the boys up, I'm moved to tears at the sight of children holding hands in a circle singing praises to Jesus, Lord of all creation; there is glory in their eyes. It saddens me to think that someday their ardor and longing may be disfigured or dimmed.

———

"Consider your grief," a friend's letter from New York instructs me. Grief: my diminished energies; my final hour. I grieve the earlier selves, the man I spent half my life hating. Sloughing through another hour on the elliptical, I grieve the burlap sack of my once sinewy glutes, the passing of my athletic body.

———

One night, I arrived home after a trip to New York in the middle hours between dinner and bedtime. Walking from the cab to the house, I could see Henry and Will sitting at the dining-room table finishing their homework. When I opened the front door, Henry ran to greet me. He had turned fourteen and was trying out, shall we say, new forms of parental affection; in any case he was not given to the old endearments. In the past year he had staked a claim to his own independence and been granted that claim. And yet there he was, my lanky beautiful teenager, with his braces gleaming like sunfish, loping toward me, the happiest man on earth. When he wrapped his arms around my waist in a tackle hug and his younger brother, proportioned at age eleven like a mini Mark Ingram, rocketed onto my shoulders, I knew this one thing with heartbreaking certainty: I would do everything I could to be the father they need, the father they love.

————

A morning, Saturday in the early spring. Soon and very soon, I will dutifully rise from my bed—I and a generation of beleaguered dads and moms—to serve once again the American youth sports industrial complex. But for a moment I savor the soft colors of the day, the sky, the still house. Life.

————

I would then dig into parenthood, focus on the work at hand, stick myself into the grind, and try to be grateful. I would resolve at least to "be fiercely in the moment"—a line I'd seen on a calling card at the vegan food market I walk to most afternoons—attuned to now, a heightened and a deeper mundane. I would become a knight of loyalty, bear the cross of ordinary surprises, nurture my children, practice patience. Guilt lingers beyond its neurotic compulsions, as it should, lest we end up tweeting drone shots of our yacht in the time of global pandemic. Even with all the consolations of *simul justus et peccator* ("both righteous and a sinner") and the old Adam, perfection was very clearly not the exception but the rule.

By "being reliably present and consistently ourselves," D. W. Winnicott writes, the parent provides a "stability which is not rigid, but alive and human, and this makes the infant feel secure." Or again: "The father may be absent, or may be very much in evidence, and these details make a tremendous difference . . . for the particular child we happen to be talking about."

———

Five years hence: My oldest child leaves for college in a week. Tonight he and some of his friends watch a zombie movie in the den, a break from packing and goodbyes. Every few minutes I hear shrieks of laughter. I drop in once and on the screen see a handsomely dressed woman in a dining room nibbling the remains of an ear; on a later walk through, a servant sits at the same table gnawing an arm bone—the well-dressed lady is nowhere in sight. It might be interesting to know that the high-school seniors huddled on the floor represent the core of Charlottesville Young Life and their leader is the bearded English major with a pillow over his face. I gather that the zombies are taking over New Zealand.

Henry is delighted, as he sits between his best friends,

their arms draped around each other's shoulders. He could be ten years old, though we had definitely come a long way from Wishbone and Raffi.

"Surely you are more stable and emotionally stronger than I," my father wrote in an email. "When we left you in B'ham, you would think that we'd just given you to Moloch to be sacrificed to the sun god, or that you had joined the foreign legion. Anyway, it's all good with Henry and our prayer is that wisdom will lead him in his journey." It's a moment that makes you wish you'd had as your father the man who grandfathers your children.

I am not more stable at all. My stomach churns like a cement mixer. I don't know about the sun gods. I do know that in such times my anxiety teams up with the approaching separation and threatens me with eternal loss.

Henry has a full ride to a private liberal arts college, and he's straining at the gates. The chirp of blue jays in the afternoon, with the sun casting long shadows on the green lawn, is today my ode to dejection. Late summer arrives with a kink in the tapestry of family life, and I don't like it.

———

And how quickly autumn comes, and goes, and winter and spring. In June, Henry and I take a trip. We are visiting the coast of Catalonia, where Bonhoeffer lived in the late '20s.

I again pose the question I've posed to my sons so often. Henry and I have been walking for hours, neither of us saying much about anything. The trail cuts sharply to the west and slopes through buckthorn and rosemary to a rocky crest. The sea below is a perfect blue. It had been six months since we were last together.

"Do you know who loves you?" I say. I assure him I would not want to turn back the clock—though if I could relive an ordinary day in the years before he left for college, I'd like that. Play catch in the field by our house. Ride our bikes to the dell. Shoot hoops. Shoot the breeze.

He rolls his eyes and claps back with his trademark snark, cutting to the quick and cracking me up.

"Well, maybe an hour would suffice," I say.

This makes him laugh. When I reach out to hug him, he's already turned back toward the main trail, and soon he will vanish beneath a canopy of wild olives.

PART V

After Analysis

Depression

It began as a lump in my throat while listening to Nick Drake's "From the Morning" on the drive to the Jackson Hole airport and ended a month later when my psychiatrist told me that, despite our best efforts to steady me through talk therapy, which had served me well for years, it was time to look elsewhere for an explanation. The signs had been there throughout the week, and for all the years preceding.

I'd come to lead a retreat on solitude and silence at Ring Lake Ranch. The first few days at the ranch had been difficult, with no phone, internet hookup, or fax machine. I felt dizzy and a little nauseated from the change in altitude. Late at night a stampede of high winds crossed the lake and pounded the windows and screen door of the cabin. I slept fitfully.

The woman who directed the ranch had once been a member of the Sisters of Charity. When she observed my haggard demeanor one morning at breakfast, she asked how I was feeling. She said it was not unusual to see guests acting a little disoriented during the first few days of a wilderness retreat. Some slept more than they had in years; others not much at all.

I told her I felt fine, and yeah, the lack of internet and cell-phone service took some getting used to, the time change too, and the altitude, and thanks for asking.

Still, she sensed something more, I could tell, and later in the day offered me the keys to an empty cabin by the main gate. I could take my books and papers there if I wanted and use the place to study.

So in the afternoons, while my wife and children went horseback riding, I drove our rented jeep two miles down a dirt road to the unoccupied cabin. Inside I spread my work on the kitchen table and organized and stacked papers beneath paper cups filled with rocks. In the short distance, a colorful meadow hummed with insects. I scribbled lines in my journal: "Today the wind blows hard off the mountains and the air is cool. The smell of cleaner in my cabin touches some ancient sadness. I stand at the desk and resolve to no one in particular that I will not be denied this sorrow."

Some afternoons I sat in a rocking chair on the porch with my well-traveled copy of Karl Barth's *The Word of God and the Word of Man* turned to a favorite passage: "Faith means seeking not noise but quiet, and letting God speak within. . . . Then begins in us, as from a seed, but an unfailing seed, the new basic something which overcomes unrighteousness." I pondered the passage, alone in the Wyoming wilderness—sometimes we must go far away from all that's familiar to hear the Word again. I wanted to know what that meant, how it might look. Slowly the afternoons in the cabin opened me to a vivid alertness, and words flowed off my pen as I sat with my Bible and Barth immersed in quietness in the second year of America's second war in Iraq. What emerged was a lament for the ensnarements of evangelicalism in a catastrophic war. The good news that God loves us unconditionally in Jesus Christ was drowned out by the sound of Jesus gunning a Humvee, storming Baghdad.

When K and the children and the rest of the group departed on Sunday, a sudden tremendous poignancy swept over me. And my decision to remain behind for a few days to write felt like a catastrophic miscalculation. A cloud front rolling in from the west covered the camp in cold shadow. But by the time I dropped my rental car at the airport, the alertness that had enabled my writing had frayed, and the poignancy was all that was left. And the Nick Drake seemed an anthem to an imminent and greater sorrow.

I'd barely been home a week when my father called with the disturbing news that a family friend had gone missing during a solo bike ride outside Atlanta. For two days, my father, then seventy-six, searched with friends and volunteers through twenty miles of dog rose and brush and midsummer heat. When he called again to say in a trembling voice that her body had been found—according to first report, "battered beyond recognition on a mound of kudzu partially hidden by vines and leaves," the brume of depression became impenetrable. Despite an impressive cast of melancholics on both sides of the family, I'd never felt anything like astral sadness filling my book of days.

I should tell you now of this fortuitous circumstance: while I'd been in psychoanalysis with Lieber, Dr. Fels, who'd saved me so long ago, had herself been training at the New York Psychoanalytic Institute. She'd left student health and set up private practice. Her office was situated on the edge of town, where Charlottesville gave way to vineyards and farmland. I happily availed myself of her ministrations. At a session shortly after my return from Wyoming, I looked up from my tears to see her own, and we both understood I had reached a new place. She had reached a new place as well. As a psychoanalyst with an MD, Fels had prescribed medicine routinely for patients coming for weekly therapy or being treated for recognizable illnesses, but she thought she had me figured out as an anxious evangelical whose symptoms remained manageable with adequate talk therapy. And then a depression hit that forced us to reckon finally with the biochemistry of my brain. I had used anti-anxiety medications in the past, but this depression overpowered those occasion-specific remedies. I'd endured plenty of depressive agitations but never taken an antidepressant. I would have most likely recoiled, had any other doctor suggested that I should.

What had I imagined as the outcome of analysis? I asked again. Not to find myself, after all these years, immured in a clinical depression. This was the rage of the illness: The realization that against my better judgement, I had expected to be healed. I loved the story my mother had so often read me from a children's Bible of the man lowered to Jesus through a hole cut in the roof of a house. I had hoped that I too would get up from my sickbed and walk. But now it was undeniable: certain fundaments of my illness would not be reached through narrative.

Trusting in the Lord with all my heart now meant accepting expert advice that I should take the pills. Fels, who I learned to my surprise had grown up in a fundamentalist Southern

Baptist church in Oklahoma, reached across the austere clinical setting to speak the words: "If you're able, I want you to borrow hope from me. Can you try to do that? I want you to accept my confidence that we can put you back together again."

Following consultations with a skilled pharmacologist at the university hospital, I was prescribed an optimal dosage of a widely used antidepressant, working up to the full strength in four to six weeks, with a supplemental anti-anxiety medication to be used judiciously under the counsel of my psychiatrist. I thought of the admonition by the French psychologist Julia Kristeva that psychoanalysts should not neglect chemistry, because in some instances, medication "can reestablish conductivity among neurons and encourage verbalization." And yes, I noted the irony that it was only some measure of health, gained through nothing but old-fashioned talk and hard analytic labor, that prevented me from being too fearful to accept the medicine.

We all know by now, I suppose, that under the inspiration of the SSRI, my brain's steely neurotransmitters released serotonin and allowed it to remain in the synapses, where it belongs—and where it enables stability—instead of being reabsorbed by the neurons. The baseline steadiness made me less subservient to irrational thoughts. The sense of helpless exposure to disturbing ideas and sounds might flare in response to patterned triggers, but their potency diminished to the point where I could fairly see the arc of their perdurance and end.

For example, the daunting, at times terrifying, prospect of writing a cradle-to-grave biography of Dietrich Bonhoeffer now gave way to garden-variety worries about research: navigating historical archives with my mediocre German skills, scoring travel grants, trying to write vivid narrative. By the time I arrived with my family in Berlin a few months later to begin a year of research, I'd emerged fully from the depression. The worry that I would not be strong enough to pull off the trip and remain intact receded into mundane concerns about passports, the children's school, and housing, all of which newly seemed like challenges that could be met.

"This must be how it feels to have a sound mind," I wrote in my journal. "Merry Christmas."

———

A complete nervous breakdown taught me nothing about God. Mental illness, benumbed and untreated, reveals the "wastes and deserts of the soul," as Virginia Woolf put it in her essay "On Being Ill." Trying to chase away madness with the petitions of faith is a recipe for disaster. There's no reason to think God wants you wasted and bare, and that is very good news. Don't believe the grace pimps when they tell you otherwise. There is no sermon in the suicide. About the church where I heard the minister say he prayed we would each have a breakdown before our thirtieth birthday? I went back to an evening service a few months ago. It was as if I'd landed in a pleasant gated community. The preacher served up his familiar tropes and meditated whimsically upon human depravity, but the words that had once imperiled me now had an altogether different effect. The sermon didn't leave me feeling injured, only bored.

The Grace
of the Strong Sin

*in Which the Anxious Evangelical Discerns
the Limitations of Psychoanalysis and,
to His Surprise, the Therapeutic Benefits
of Martin Luther's Doctrine of Sin*

The thought first came to me in the churchyard—I'm on an island in the Baltic Sea, where I'd rented a room in a beautifully restored eighteenth-century guest house an easy walk from a sixteenth-century church. Well, it wasn't a *thought* I had in the churchyard, standing among the worn markers of men and women who'd lived and loved and died five hundred years before, so much as a sense of the importance of just getting on with it—of not living perpetually trapped in a room with desire and freedom and trust in God and terror of myself. I had to venture forward.

A week later, I'm back in Berlin. I'm sitting at Café Entweder-Oder, on a street in Prenzlauer Berg—not too far, in fact, from the rooms Bonhoeffer lived in after he returned from New York in 1931. Sunlight splashes through the linden trees—"the trees of lovers" according to Germanic folklore— onto my notebook. In my bag is Kenneth Kirk's *The Vision of God*, lectures from the 1920s about the means Christianity proffers for apprehending God and infinite love. It's a forgotten gem, this book, with its stylized avian cover (I now realize I'd lifted this particular copy from the library at Ring Lake Ranch).

Earlier, working on the biography, I'd returned to the passage from Bonhoeffer's prison writings where he tells Eberhard Bethge that he is only reading the Old Testament and that he is relieved, delighted in fact, to find passions lived and embodied, men and women feeling and lying and fornicating for the glory of God. There, in prison, Bonhoeffer wrote of his regret that he would not be able, in his lifetime, to enjoy, to give himself to, to accept the pleasures of the body. If he could do it again, he told Eberhard, he would accept the grace of the strong sin. (He also wrote of the heat in Berlin—that he was writing naked but for a pair of silk boxer shorts he'd actually purchased *for* Eberhard but had managed to bring into Tegel.)

I considered retrieving the Kirk from my bag. And it was then that I realized that, for all the good analysis had done me, it had not given me permission to sin. It brought me to knowledge of the freedom to choose, but balked when I wanted to choose the sin. Which meant that the benefits of analysis— creative growth and self-respect, which had concrete positive results, familial and professional—collided eventually with the persisting desire to do the feared thing. The apocalypse of sexual transgression remained unchallenged at the level of lived experience, and so I continued to think that impurity would bring my ruin. I was still the fifteen-year-old boy who wrote in his journal: "Tribulation—earthquakes—hailstorms—110-pound blocks of ice—sun will go dark—moon won't shine. Satan wants me as his instrument."

How could I know I was free from my primal tyrannies if I had not ventured the strong sin—tasted it, smelled it, jarbled my face with its juice and gleet? That is to say, how can one know God's forgiveness if one never ventures to the territory where forgiveness is most desperately needed—and perhaps most warranted? This is where psychoanalysis needs theology, or needs, perhaps, prayer. Analysis, because its own repertoire is lacking, doesn't enable the analysand to feel and experience the bodily effects and reassurances of forgiveness. To find forgiveness, you must first put yourself into the place you've always associated with terror and alienation and there feel God.

I had psychoanalyzed my desire down to the bone. But the work of understanding the structure of my desires and turning the understanding into story kept running into the same dead end: my body did not trust the mind's achievements. What I suddenly understood at that Berlin café was that analysis had made it possible for me to place myself there—to take hold of the strong sin—but analysis was not a replacement for doing so.

Enter Martin Luther, the trash-talking Reformer, bloated and flatulent, fomenting then denouncing revolution, who stooped to our beleaguered analysand and said, "I've got a few things to say about this." Martin Luther was the German monk who remade the Christian consciousness of the West when, on the Eve of All Saints 1517, he nailed a parchment list of Ninety-Five Theses in Latin to the wooden door of the Castle

Church in Wittenberg. Meditating on St. Paul's Epistle to the Romans—"The righteous will live by faith"—Luther found a modicum of relief from the torments of a guilty conscience. "Faith alone," *sola fide*, saves sinners bound for hell. "Faith alone" proclaims men and women righteous by the blood of Jesus. Meditating on this, Luther said he felt as if he had been led "through open gates into Paradise itself."

Divine grace, Luther proclaimed, was an "alien righteousness," *extra nos*, the gift that comes to the sinner without conditions from the one who is wholly other. Such righteousness, Luther proclaimed, is received in two ways: by faith once for all and then in a manner infused gradually into the will, efficacious quite apart from the individual moral agent. I've never understood how this works. It seems a fantastical act of divine puppeteering if sin is not only what you do, but who you are.

And yet meditating upon Luther's words on this fine morning in Prenzlauer Berg inspires the thought that God wants of me nothing less than a sailor dive into his tender mercies.

Luther made the case as a hypothetical—"That, if we have no such thing as Freedom of Will in our own things and works, much less have we any such thing in divine things and works"—only to return all agency and merit to God's transcendent freedom. That's the theological axiom.

But at the pastoral level, Luther knew more than he let on about the conditions of Christian freedom. I once heard

---

The part about *nailing* is disputed, I know (there's also the pot of glue hypothesis, in which the Theses were not hammered but stuck to the door with a dab of *Klebstoff*), but such quibbles have rarely inhibited the fantastical reports of the faith of our fathers.

a lecturer tell the story of Luther mentoring a certain monk. The young man came to Luther every week—or maybe it was every day—with the same confession: his daily path led him by the home of a pretty peasant girl and he could think of nothing but bedding her. Luther finally told him just to go ahead and sleep with the woman—for there was no better way to leave the Devil quaking in his boots than to bank recklessly on grace. Luther quickly added, at least this is what I remember the lecturer saying, that as much as he would have wished himself in the monk's stead, he—considering his fragile nerves and obsessive mindfulness of hell—would definitely not be able to pull the venture off. But he wished the monk well and would be (presumably) standing by for details.

If your will has been broken, if you remain enslaved to fears of disintegration, you will need to experience the grace of the strong sin. This is not antinomianism; it's a stroke of therapeutic genius. Without the freedom to sin, it is mere sophistry to attribute virtue to inaction.

The fear of God may be the beginning of wisdom, but fear of the body calcified into a certain spiritual paralysis quashes the confidence of the gospel. You can't smile at Satan's rage if you're cowering on the mourner's bench clutching Philippians 4:8 as if it were an amulet to avert the gods. I don't think it's off the books to say the strong sin may then serve as a propaedeutic to joy.

"Christians talk about the horror of sin, but they have overlooked something," said Binx Bolling, the New Orleans stockbroker and self-diagnosed "malaisian" in Walker Percy's *The Moviegoer.* "They keep talking as if everyone were a great sinner, when the truth is that nowadays one is hardly up to it. There is very little sin in the depths of the malaise. The highest moment of a malaisian's life can be that moment when he manages to sin like a proper human."

I'd run up against the rigorist pieties of the couch. The analyst cannot say, "Indulge your heart and have the woman," even if it's obvious that's the thing the neurotic needs most. The analyst can't say, "Sin, and sin boldly." But you know who can? The God who puts the sinner as far from his sin is as east is from west. The God on whom Bonhoeffer's sights are fixed when he writes that in the Old Testament "men tell lies vigorously . . . kill, deceive, rob, divorce, and even fornicate (see the genealogy of Jesus), doubt, blaspheme, and curse," often "to the glory of God."

———

Ten months before his arrest, Bonhoeffer wrote in a letter to Eberhard that he felt as if he were "on the verge of some kind of breakthrough." He newly saw that the disciple was connected even more deeply, more intimately to the physical world than he'd grasped before, and he, Bonhoeffer, now felt newly open to "the worldly [*weltlich*] realm." That openness in turn deepened his engagement with scripture: "I am living, and can live, for days without the Bible," he wrote. But when he returned to the Bible, he could hear and experience the "new and delightful . . . as never before."

In prison, Bonhoeffer turned to genres other than theology, sometimes as vehicles for his new thoughts, sometimes for the challenge of playing with language in ways that had hitherto felt frivolous. The results in fiction and drama were uneven. I think of them as beautiful failures. His poetry fared somewhat better; some of his late verses are now sung in Lutheran churches. But in these efforts, we see him moving toward an entirely new apprehension of the Christian body, of pleasure and desire.

On a hot July afternoon, he sat in his cell wearing a pair of gym shorts and a dress shirt and wrote to Eberhard about the "concrete bodily experience of heat" that was overpowering, that provoked his "animal existence," "his corporeal being," with a singular urgency—"not just to see the sun and sip at it a little, but to experience it bodily." He longed "to feel again the potencies of the sun," he said, and he remembered past summers: with Klaus in Rome and Libya in 1923; the first trip with Eberhard to Naples in 1936, the colors of the sun hitting the Mediterranean as observed on the ferry to Majorca; Christmas 1931 in Cuba. He pondered "the romantic enthusiasm for the sun," the cult of the sun gods. And then suddenly he asked: "Does it make Goethe or Napoleon a sinner to say that they weren't always faithful husbands?" And he wrote, then, of strong sins, sins dared for the purpose of "nurturing intimacy with others." (Later, when I read Rowan Williams, in "The Body's Grace," limn the ways even sex glossed by the tradition as transgressive could bear holiness and humanness to its practitioners, I would think of Bonhoeffer; I would anticipate the conversations the two greats would have some day or, given what Williams likely believes about saints, are having even now.)

Perhaps one—only one—of Bonhoeffer's achievements here is to give us a glimpse of Luther, or Luther's Christianity, cured; Luther restored; Luther freed from inner torment.

———

Grace must be felt, and forgiveness embodied. Analysis cannot feel theology for you. It can't substitute for your testing and receiving whatever there is to be tested in and received from that queen of the sciences. This addresses an aporia in the tradition, perhaps: it is only when our hero survives the strong sin that he knows, in his body, that God is real and God is forgiving.

And so one afternoon I take the train to Wittenberg, and there, in the shadows of the Castle Church, I resolve that if the Spirit should ever lead me, to the place of the strong sin, I would go the distance and trust that grace will bring me home.

On Christian Counseling

I was primed, I suppose, to evangelize—and yet I was also
primed for secrecy, perhaps shame. That is, because I was
agog with enthusiasm for all that I found in psychoanalysis, I
wanted to spread to everyone the good news and urge them
to hie themselves hither. And, because of a subterranean
instinct for privacy, the only person to whom I spoke this
thrall was K. Which means she heard a great deal of rapture
and a great deal of urging—she should be in analysis, too, I
thought; she should involve herself in this work that I found
so liberating.

I could see that my enthusiasm was an aggravation, but I
couldn't rein it in. She and I together—with the aid of skilled
analysts—would cast off the tyrannies of our unsparing gods
and return home every night from our different couches with
new insights to share. I proselytized thus because of the growth
analysis had enabled, because I loved K, and because I could
see that she too was broken by her religious upbringing—as
burdened by expectations of perfection as I was.

"Analysis!" I'd say.

And she would reply, sometimes calmly, sometimes with heat, "Your story is not my story; your problems are not my problems."

Classic resistance, I'd think to myself.

And so I kept at it, and in time, perhaps just to silence her husband, K arranged a consultation with an analyst who worked out of her home in Pikesville (though it might be better to say she worked out of the cluttered dining room of a squat brick house with low ceilings). I'd understood from Lieber that the analyst was a Russian Jew who'd been educated in Paris and analyzed by a Freud epigone. This sounded amazing to me, but it did not make for a good fit for K—who, in the most literal sense, could not understand much of anything the analyst said. She spoke quickly and with a thick eastern European accent and asked K few questions, as far as K could tell.

And then, some years later, K found her way to Sonja, whose practice answered to the label "Christian counseling." Sonja had been praised by a friend we much respected, and I even heard a local indie rock band give Sonja a shout-out during a show. Yet I had my suspicions, of course.

I own a small library of books that form the canon of the Christian counseling movement, and although I'm not a doctrinaire Freudian (nor are any of the analysts I know), the militant hostility I find there to Freud's most self-evident insights—the unconscious as a reservoir of aggression, desire, and memory; the tyranny of the superego; the necessity of proper self-love—did not inspire confidence in Sonja or her ilk. The very idea of "Christian counseling" seems to me a synecdoche of my childhood evangelicalism's contempt for the secular mind; it recalls the constraints I wanted both K and me to be freed from.

And so at first K and I spoke only gingerly about what she was finding in Sonja's company (and to this day I think it's a delicate matter what you share with your partner when you return home from the therapist's den). Over time, though, she was willing to say, and I was willing to hear, how Sonja helped her articulate her own "I," how, especially, Sonja's questions and gaze helped K untangle her own self-understanding, and her own theological self-understanding, from her parents' (who insisted that, despite their oldest child's manifold virtues, neither she nor her husband had fully grasped or yielded their life to "the spirit of Jesus").

Through K, I discovered that there exists within the Christian counseling world an eclecticism not envisioned by the "nouthetic" biblicists. That adjective is derived—according to Christian counseling guru Jay Adams (remember him and his dedication to breaking the child's will?)—from the Greek New Testament words *nouthesis* and *noutheteo*, which Adams translates as, respectively, "confrontation" and "to confront." Adams's touchstone verses include Romans 15:14 and Colossians 1:28, and in his rendering, the apostle Paul teaches the church at Colossae that the Jesus we proclaim "confront[s] every man nouthetically"—which sounds like abject torture. And a recipe for disaster. And what I'd been subjected to most of my growing-up years—admonitions delivered with the presumed authority of scripture, supposedly unblinkered by self-interest, that I should receive in turn as the path to wholeness.

Which is not to say I'd ever put myself in the clinical care of someone who claimed the moniker "Christian counselor." I'd be reluctant to enter a therapy session in which I felt doctrinal commitments would even indirectly press upon me.

As I write that sentence, of course, I can see the worry about mixing of doctrine and therapy is more likely generated by my own habit of mind than by any impulse of Sonja's to throw a scripture verse into the mix. Yet doctrine's pressures can too easily be transferred into a Christian discourse on obedience and faithfulness, and—whatever the origin of that doctrinal hum, within or without—the insinuation of doctrine into clinical space would preclude exploration of too many regions of subjective life—of desire, aggression, and inhibition, for example.

The narrative of self-becoming depends on a freedom given the analysand in the psychoanalytic dialogue, and that particular freedom—a freedom of association limited only by the length of session and the analysand's willingness to find salience in both the extraordinary and the mundane—is manifestly not part of the Christian moral vocabulary; even in the confessional, after all, one's revelation is already narratively framed. I came to therapeutic work carrying boatloads of theologically glossed evasions, one Christian frame after another telling me I couldn't talk about this or that. The last thing I needed was a "Christian counselor" who either would, or would be thought by me, to supply the same.

Still, there were evenings when I found myself envying K. Her sessions with Sonja weren't only, or always, spiritually edifying, but it soon became clear that spiritual edification—not as instruction or dogma, but as the creation of a space where K's soul could be nourished—was sometimes available in Sonja's ministrations. I am sure, of course, that my own changing perceptions of Christian counseling had little to do with what actually went on between Sonja and K—that, as ever, if my fantasies of K's therapeutic hour offered any insight, it was only insight into myself. Ergo, I came to see Sonja's office not as doctrinally inflected narrowness, but as something more capacious, and when I daydreamed my way into that office, I imagined that I might come away from Sonja with a clearer understanding of my spiritual hopes and longings. I thought I might feel more of a sense of purpose. I thought that I might perhaps feel, sometimes, guidance.

I wanted, sometimes, to be guided, to be built up—but that's just not what analysis does. It does what I needed it to do. I needed to sort through a discrete set of conflicts—the dissolution of trust, the warring of spontaneity and shame, the remnants of terror buried in an "utterly corporeal body" (to quote Teresa Brennan's book *The Transmission of Affect*)—and I needed the confidence that I could do so while holding steady. Or maybe the better way to put this is: from analysis I needed reassurance of the capabilities of human being and body and mind, and with that kind of reassurance I could begin again to ponder God, because with that kind of direction I could dream and imagine.

One night—what was K talking about? Something about how Sonja will often ask after her religious well-being, encouraging openness to the Spirit's presence in the therapeutic work. Sonja will ask K whether she's finding her way into a spirituality that feels her own, free of scripted piety, of the expectation to use the right words, to speak the right language. I found myself thinking that her counseling with Sonja contained a kind of benediction, and then I wondered, *What do you call the experience of prayer where you are given the freedom of complete openness?*

In 1935, Bonhoeffer began directing a seminary in the Pomeranian village of Finkenwalde, where pastors could be trained outside of the university theology faculties (which were now fully aligned with Hitler). Even before the Nazi takeover of the university, Bonhoeffer had lamented the Teutonic entrapments of German theological education, which he found wintry, confining, asphyxiating. Finkenwalde was an experiment in creating winter's opposite—a theology "of spring and summer."

Bonhoeffer taught ecclesiology and biblical interpretation, to be sure, but his deeper goal was to create a community shaped by the Sermon on the Mount, a community of "uncompromising discipleship," of forgiveness and charity and prayer (and also, it should be noted, a community of play and swimming and tennis, of leisure and forest walks and joy). A student once said to Bonhoeffer that he found it impossible to pray for even ten minutes, never mind an hour. Bonhoeffer reassured him: "That's as it should be; you should not feel ashamed or unworthy, but allow yourself the freedom to have any thought or any concern and to accept those concerns as part of prayer's grace."

To the extent that I have such freedom, I found it in, or via, analysis. But it was only in the afterglow of one of K's visits to Sonja that I connected my analytically grasped freedom with the freedom Bonhoeffer named when he spoke of address to the Lord. Mostly, these days, when I think of K's sojourn into Christian counseling, I just give thanks for the space she found there. But sometimes, still, the thanks are touched with envy.

PART VI

Quiet Days in Charlottesville

Oh, Merton

Reading Thomas Merton this morning, I'm irritated by his ruminations on the "bare flight of time." Time's "bare flight" is the single aspect of an exterior life about which the monk should not write. "Much more important are the events that take place in the depth of a monk's soul," Merton says. Monastic time—the liturgy of the hours, the fire watch, the daily tasks around the compound—is pure time. The kind of time that doesn't "fly bare"—or, rather, whose flight is transfigured by God's response to our oblations.

Well, fine. My three children have now entered the kitchen wanting four different things. My oldest is unhappy with his haircut—which looks exactly like that of every boy in his peer group. I suggest that he call the barber shop and ask Mr. Kim if he's got time to take a little more off his bangs. My middle child can't find his iPod and says his older brother probably stole it, to which the older brother replies as he's leaving the room, "You're such an idiot!" The youngest, my daughter, needs a ride to violin lessons and help finding an earring. I offer reassurances to her and to myself as well—my wife's yoga class will be over soon.

For a moment the room is quiet, and I turn back to *The Sign of Jonas*. Until the two brothers reappear in the doorway, best of friends, and ask for a ride to GameStop.

I return to Merton because I still love him. But really, Merton doesn't have a clue. How fine the Cistercian disciplines that absolve him from the tedious necessity of making plans and of coming to many personal decisions. Oh, that's deep, man. This keening Kentucky kenosis of fire and night. Mastering the silent deep where it's real, real, real.

I have an idea, brother. Follow me into the limpid chaos of family. Try this on for a while, if you want truly to feel the tensile strain of self-denial; try grinding it out with the same woman for twenty, thirty, forty years. I know that heroism all too well, and I know the hankering for the one pure thing. I don't begrudge the boho monks their flight to the abbey. But I'll bet you a quick roll in the cornfield that a monk's life is a drink with jam and bread compared to all the crap my wife and I have to do in the next twelve hours.

———

Sunday morning. K has taken our daughter to church. I've made a pot of coffee and spread the *New York Times* out on the kitchen table. My two sons are asleep upstairs. I told K I'd take them to an evening service somewhere. After all the shit their schools, coaches, music teachers, Young Life leaders, and parents put them through Monday through Saturday, they deserve a lazy morning. My dog naps with his head resting against my left leg. The house is still.

Two months ago these arrangements would have made me uneasy. I'd come to dread Sunday mornings. The hurried showers and half-eaten breakfasts, the minivan hell of raw nerves and bad breath, and we're always late anyway. And then the ordeal of the sermon. And then the mad rush to the parking lot after (or before) the benediction, with more grumblings from the tribe, mostly mine, over the thing that has just transpired.

I'm probably not the only Christian who has sandwiched Nietzsche's *Antichrist* between the hymnal and the bulletin to survive the Sunday service. My copy dog-ears right over my favorite pericope. Please turn with me, if you will, to page 106, and feel free to follow along in your own pew Nietzsches—or simply listen with an expectant heart. My Ludovici translation, in a two-tone, easy-in-the-hands paperback, opens to his comfortable words: "The Christian Church is the greatest of all imaginable corruptions. The Christian Church leaves nothing [untouched] by its venom. The Christian Church turns strength into shame, vitality into listlessness, which it then calls a virtue. The Christian Church sucks all the blood and all the love and all the hope out of life."

Memoirist Kathleen Norris speaks of sermons as "word bombardments," "agony for a poet." And no picnic for the rest of us either.

Some of the craziest things I've heard in my life I've heard in sermons. One preacher gushed over his new Glock and the thrill of speeding in his car. Another proclaimed the gospel of the "thug Christ," and said you should never worship a Jesus you could beat up.

Death-row inmates are given more license to talk back.

Listening to a sermon, you can't raise your hand with a question. You best not shake your head in dismay. It's an experience rarely observed outside of totalitarian states—which may be one of the reasons some churches have experimented with "sermon talkbacks" after the service ends; but by that point in the morning, having survived the ordeal, you're more likely to make haste to the parking lot.

A modest proposal to the editors of the DSM: in your next edition, please consider including among phobic disorders the diagnosis homilophobia. "Someone suffering from this condition can expect to experience a very high amount of anxiety from merely thinking about sermons, let alone actually hearing them."

This is why I was not eager to join K and Nan at church.

In time, the boys wake up. They want Cajun pancakes—pancakes smothered in salted butter, confectioners' sugar, and warmed vanilla—and the smoked bacon I buy from a former student who quit grad school and launched a sustainable, pasture-based farm in Culpeper. We'll have three newspaper sections spread out on the table. Henry pulls out the sports section, and Will pulls out the arts review, while I read Maureen Dowd on W. There is always music playing. Bach on Sunday mornings, mostly, though sometimes Switchfoot and R.E.M.

I'm serious about taking them to a new start-up church with an early afternoon service—the one with the whimsical untucked pastor Dave, who plays video clips of *The Office* and *Star Wars*. I want them to hear the language of the gospel, the *evangelion*. The wager that the world would be more peaceful if all children knew deep in their bones that "God so loved the world" and that "Whoever lives in love, lives in God" seems reasonable.

But, alas, I lose track of time, and it's suddenly too late to go. Or maybe I know exactly what time it is.

In any case, to cheers of approval from the playroom, I announce, "We're doing home church today. Huddle down-stairs in ten." Soon my congregation of pubescent brothers in mismatched pajamas reads Psalm 8 aloud from the popular paraphrase *The Message*. From their mouths, still flecked with powdered sugar, I hear again how God made humankind but a little lower than the angels, "bright with Eden's dawn light," fashioned the heavens, "dark and enormous," as "handmade sky-jewelry" sparkling gloriously for our delight, and filled the earth with sheep and cattle and "animals out in the wild," with birds and fish and "whales singing in the ocean deeps."

Then I tell the story they've heard many times before about Fannie Lou Hamer, of Ruleville, Mississippi, who dropped her satchel in the cotton fields of Sunflower County when the Pen-tecostal fires of the civil rights movement swept through the South and went to work for Jesus and disenfranchised Blacks in the Mississippi Delta. Then—fifteen minutes on—I play a track on the turntable by the Dixie Hummingbirds, "Have a Little Talk with Jesus," after which I reach out to the boys for a bear-hug benediction that quickly turns into a wrestling match.

I'm grateful for the Sunday hours and crisp morning light. The sun cuts clean lines through the trees. A storm that had been hanging over the Shenandoah Valley all weekend veered westward and chased away the sticky heat.

———

I have now rounded fifty, and fifty-two, and fifty-five, and as for my once athletic body, I've made my peace with the elliptical machine. Until fifty, I'd felt pretty much the same as my thirty-something self. Admittedly, some years ago, on the basketball court my defensive skills abandoned me. But I didn't care. I'd hated lateral drills in high school even more than suicides. But with quick hands and speed, I could frustrate most of opposing point guards in Alabama 4A (now 6A) competition. Ask my father about my basketball career, and the first thing he'll tell you is the time I took a wide-open jumper—so it looked to me—from the top of the key, and a sixteen-year-old Chuck Person slapped the ball back so hard it rolled out the gym door into the concession area. But I ran a solid point guard, averaged 10 points and 8 assists a game, and held my own with college players in pickup games.

In my early forties, I more than made up in three-point shots what I'd lost on defense. I could still run with the kids at North Grounds; and on Sunday afternoons, I met up with a group of old jocks at a rec center and ran full-court games for two hours.

By the time I bowed out of the Real Men League—real name—I was spending as much time in the pregame fastening of braces and straps as running the court; and I did not go gently into late middle age. My aging, aching body, "the grind of bone and socket" (to recall B. H. Fairchild's elegy for the over-the-hill baller), pissed me off. I came to hate for two hours each week everyone else on the court. I can't think of many things more pathetic than the sight of old men all up in each other's faces over a disputed foul. My friend Karl used to say that an aging man's denial of death inspired the delusion that with a little more gym time, he could still improve his vertical jump. Guilty as charged.

As I neared fifty, I remained under the impression that I could not only improve my jump, but that, with willpower and the right vitamin supplements, I could match and maybe even better my best time in the 440-yard run (a 50.3, clocked in the twelfth grade), even though I was already beginning to feel my lower lumbar region firing sciatic darts into my left foot that would lead to a spinal fusion.

Still, when another birthday arrived, I could reassure myself that beyond me lies thirty or thirty-five good years, barring death, disease, or the end of the world. Good things can be accomplished. Colossus-brain Immanuel Kant began his three critiques at the age of fifty-seven: *The Critique of Pure Reason*, *The Critique of Practical Reason*, *The Critique of Judgment*. The last, for two centuries overshadowed by the first two, yet now considered by most theologians to be his best, dropped just short of his sixty-fifth birthday. The world of the mind would never be the same. (Not that anyone need give a dagnabbit about the crowning achievements of a xenophobic old Prussian, or be impressed that I kind of do.)

Walker Percy was forty-six when his first novel, *The Moviegoer*, upset *Catch-22* for the 1962 National Book Award, and fifty when he finished *The Last Gentleman*, the dreamy account of Will Barrett, a displaced Southerner who suffers from a nervous condition—my favorite of Percy's fiction. Of course, I missed those marks, but I'm still on track with Percy's *The Second Coming*, which picks up Will's story at middle age as he suffers the slings and arrows of domestic unhappiness in a claustrophobic southern town.

But I'm not sure what to think of the man I'm becoming. I wish I could wrap my arms around the remaining time and say, "Do you know who loves you?" But you cannot trust time.

———

A few days later, with Nan safely transported to music lessons and arrangements made for a "play date" (the preferred locution, in those years, of parents of a certain demographic), and with the two boys bivouacked somewhere safe, I hoped, K and I collapsed into bed. It was three o'clock in the afternoon.

"I am so tired," she said. "It's like I've got a kind of illness. I should take more vitamins."

"You should try Allegra-D," I said. "Or maybe four Sudafeds, since you don't have allergies. Works best with a second cup of coffee."

We laughed, held each other, and feel asleep. A half hour later, K startled awake with an "Oh, crap," remembering that Will needed a ride home from driver's ed. I'm late for a meeting at the university, I tell her, which is true. But when the front door slams and K's minivan peels out of the driveway, I think maybe it's too late to go at all and text my regrets to the grants and fellowships committee (bad sinus headache; I'll send my rankings in an email). Then I roll back into bed in the quiet, feeling a little guilty that I've misled (lied to) my colleagues and, sort of, my wife. But not that guilty.

———

Oh, Merton: "I wish nobody had ever told me it was a good thing to attempt to know myself."

Here we are again. I return for the heat at the heart of the divine mystery, for the way he goes to the extremes of things and feels free to imagine how it would be if he stayed there. But sometimes I can't see a meaningful difference between his gleaming monasticism and the melancholy piety of decrease that taught me early on to distrust even the small pleasures. To wit: "The more you try to avoid suffering, the more you suffer. The one who does most to avoid suffering is, in the end, the one who suffers most."

Oh dear Father Louis: please go sit in the truck.

Quiet Days in Charlottesville

It's a clear Sunday morning in Charlottesville. Ginger sleeps on the front porch with her head on the cool brick. In his strange little book *The Christian and Anxiety*, Hans Urs von Balthasar writes about the anxiety that comes upon Jesus in waves: first an initial shudder at the grave of Lazarus as he brushes against the world of the dead and soon-to-be-unsealed abyss and darkness, then a new shudder in the temple at what is now certain, and then others on the Mount of Olives, in the garden, and on the cross. Balthasar calls the experience a "bracket."

It was either some unconscious awareness of that bracket or a Spirit-enabled memory that kept me from Quentin's fate.

The bracket is a consequence of love. Love, imperfectly rendered or overexerted, is finally greater than the hurts it exacts. The solicitudes of love that feel heavy, intrusive. The scars that reveal love's imperfect constancy and sometimes heal.

A clumsy term, no? Clumsy, and of course he's riffing on the whole phenomenological arcanum that coursed through continental philosophy a century ago and especially excited the Catholics. But a bracket is also something that supports; it is a small shelf on which useless beautiful decoratives are placed. It is also the punctuation that encloses a noted word or phrase and pulls it out of its context. Maybe all that is what Balthasar meant too. A support. A distinguishing from your context, your ambient all-the-time, your clutter of other words, words, words.

⸻

Watching family movies on a recent visit with my parents, who, in their late eighties, still live in their modest two-story home in northwest Atlanta, nourished by the kindness of neighbors and former parishioners, made me think of St. Augustine, how he marveled over the passing of time—the question of whether time could be perceived, or measured, after it had passed, *when it was not*—and questioned the relationship of past, present, and future: "Perhaps it might be said that there are not three times, past, present and future, as we learnt in boyhood" he wrote in the *Confessions*, " . . . but only present, because the other two do not exist. Or perhaps that these two do exist, but that time comes forth from some secret place when from future it becomes present, and departs into some secret place when from present it becomes past."

Future and past cross the threshold of eternity in a luminous present, before vanishing into nonbeing. But nothing is lost to memory.

The movies are a single three-hour compilation of our

family life, 1958 to 1976, the year of my birth to the summer I left for college. A shoebox of reel-to-reel tapes transferred to a DVD in random order. Like all home movies, not a compendium of facts, but a deluge of impressions.

In my father's Goddardesque production, I'm a toddler crawling resolutely toward my mother in our tiny back yard in Mobile. Ivy skeins around the slatted fence sparks against the porcelain Madonna next door weathering the heat in her ever blue robe. A crew-cut country boy walking down a parched dirt road lobs a rock at the camera (good God). A gangly teen in cutoff jeans jackknifes into a swimming pool—is this the Buena Vista Hotel in Biloxi?—soaking his great-aunt Emily, who'd been napping in a cabana, a *Ladies' Home Journal* on her lap.

Cut to a black bear approaching a gaggle of tourists somewhere in the Smoky Mountains, and the swirl of asphalt and forest, as my father hauls ass back to the car.

London in the liquid light of a July morning, and I'm flashing a peace sign. I'm off to find the Don McLean concert in Hyde Park, I remember; I never found it.

Forward to the melancholy blues of my mother and me in a train cabin on our way to a Christian commune in the Swiss Alps. I look directly at the camera, then out the window, then back to my father. What had I seen? The flutter and dart? I'm trying to remember. A brutal industrial zone carved into a mountainside? The Vevey funicular vanishing into fog?

It's the rare case in which I've seen myself as anxious subject, and I am moved to tears. I wish I could tell this fifteen-year-old things are going to be okay. That the anxiety he was feeling then—because that's what it was—will not let him float out of his body for a dissociative view of his suffering. It blocks perspective, and he'll need to figure it out to become unblocked, but he will.

Time dissolves in a vertical burn, as Waukaway Springs— *ah, Waukaway*—a cold-spring swimming hole north of Laurel, flashes bright. In the June heat, I stand at the jukebox with three other church boys, feeding a quarter into the machine. How different the instant would have been had I felt the body's grace. To love the sunlight and breathe deep. But there's the fear as I face the camera. Busted.

———

I've spread my papers and books on the dining-room table alongside last night's Thai takeout containers and a few empty Warsteiners. The sight of this would drive my wife nuts. But she's out of town for the week, so here I am flying high at my own private Zeppelin hotel party.

The day ricochets between scribblings in my notebooks, rearranging my library, and buying things online I don't need. Living the dream.

Most days it hurts to write, to sit down at a desk—or to stand, though I own two adjustable-height desks. My left leg hisses from a half-successful spinal fusion; my left hand is palsied and numb at the carpal nerve and needs surgery. My ass hurts.

Last week, I was teaching Kierkegaard, who thrills me with his leaps into the dark nights of resignation, but who leaves me wanting hope. His absurd remains an idea that winnows away the other. Better to believe on the strength of the beloved, on the strength of the cloud of witnesses, who saw a light in the darkness and hoped first and foremost for the redemption of the world, who hoped for the redemption of the world for the sake of the other.

I find myself once again writing about the civil rights movement. It seems to me, these days, that the student volunteers whom I've so often adoringly hymned—those who came South to help out with voter registration campaigns, freedom schools, and all that—were after more than bringing southern segregation to an end. There they stand, a band of sisters and brothers lost in ecstasy, arms raised in praise, blinded by the resplendence of the moment, tears pouring down their faces, the uniforms of denim soaked with sweat. You'd never know what dangers lurk in the southern summer outside. Their singing has sealed the room against the night. What were they looking for? Yes, they wanted to build the beloved community. But more: they longed to disappear. They were casting about for an exit—an exit from their lives before the movement or, for some, an exit from their commitment to the movement, from the narcosis of high-minded resolve.

This, perhaps, is part of the work those students did for me, all those hours I devoted, in my Baltimore cathedral office, to reading them, to chronicling them. As I wrote about voter registration, marches, and boycotts, my thoughts searched for an escape of my own. I imagined other lives for myself—an intentional community built in a pecan grove where the sun is autumn warm and tenderness holds the day—there I would feel free. I would gather with friends, with my soul brothers and sisters, at nightfall and admire the changing light, the adding and subtracting of color and shadow, the burst and glow and bluing of the setting sun.

Perhaps the movement—the sit-ins, marches, freedom rides—was not all about hustle and terror—and perhaps I wasn't betraying the cause in my drifting attention. The condition for achieving beloved community was a certain kind of stillness in the midst of frenzy and noise, learning to move at a different pace. It delighted me to learn that more than a few student volunteers left the South out of boredom.

Jane Stembridge, a white minister's daughter who spent all her youthful energy working for Black suffrage, described waiting as a discipline, as a spiritual and aesthetic attunement to the "lonely center of the spinning earth." And in her opinion, the movement could have used a lot more frivolity, beauty, and play. She wrote in one of her remarkable Mississippi poems:

When we loved,
we didn't love right.
The mornings weren't funny,
and we lost too much sleep.
I wish we could do it all again
with clown hats on.

Most who pursued a movement life, I now see, understood the quality of their life together as an attempt to live into a new and distinct kind of time. Let's call it the gentle urgency of now.

———

Again I sit at the window of my dining room on a Saturday morning, a weekend in middle spring. Ginger lies at my feet, snoring, inside her Zen cone. I can't get her to stop gnawing her hot spots, so on goes the cone. She doesn't seem to mind for now.

A tree flares in the morning light. What kind of tree? I don't know. The tree in my direct line of sight that I've looked at a thousand mornings and that blossoms even in the winter and that I haven't tried to identify. Crimson reds fountain forth beneath the green.

And yet, a repeat performance by this body, this brain; an arsenal of dark surprises, my hushed temple of dread? I could so easily make a case for oblivion. This distrust of the body, this distrust now in its seventh decade, I seem to have made my book of days. All that appears to be good in this world wears out in time. Is this a morbid sadness?

And yet I am a Christian, and no less obliged now in these late days to profess my belief in creation and redemption, in the sustenance of grace, in the goddamn resurrection of the dead.

———

At a wedding, I notice a prayer I've never heard, or at least listened to, before: "The union of husband and wife in heart, body, and mind is intended by God for their mutual joy; for the help and comfort given one another in prosperity and adversity; and, when it is God's will, for the procreation of children and their nurture in the knowledge and love of the Lord."

What do I want for my children—what have I always wanted for them? I wanted them to have happy and secure childhoods, to grow into an expansive vision of human possibility. I wanted them to receive disappointment when it came, as ingredient in the mung of things. And, yes, I wanted them to be raised in the language of the Christian faith, to experience the richness that is there—that I know is there—beyond its conspicuous profanations. I wanted them to know that reason, justice, humanity, and culture find purpose and power in their origin, which I, as a Christian, affirm to be Jesus Christ. I wanted to relieve them of the pressures that I had felt as a child and a young man. I wanted for them the freedom to move more capaciously within a large field, an expansive field of grace, though I knew that such relief came with risk.

Of course, I've not only wanted Will and Henry and Nan to *have* those things—the happiness, the ordinary unhappiness, the relief. I've wanted their happiness and freedom to be *inheritances*, to be a gift from their father. Perhaps, like all gifts, they'd be traced with damage, the damage of my own incapacity to give freely, to give a gift untangled from my desire to be the one doing the giving. Or perhaps that is not a damage; perhaps it is perfectly the office of a father to desire to give such gifts. And anyway, if Christianity shows us anything, it is that only God gives perfectly.

They are now grown, these children of mine. And I believe

perhaps I did give them those capacities, for ordinary unhappiness, and ordinary, and, sometimes, uncustomable joy.

Hiking with Henry a few years ago in Spain, I told him that I had regrets as a parent. I told him that I regretted not having talked more about God. "I regret that I didn't give you a fuller sense of how much I love the Bible," I said, "how much I love the rich Christian inheritance."

Henry slowed to a promenade. "You talked about that all the time," said Henry. "You read to us from the Bible and the daily *Losungen*, we lived with you while you were living Bonhoeffer—and went with you on research trips to Barcelona and Rome and London; you told us stories about Dorothy Day and John Lewis and granddaddy Bob"—my father, who had preached "Amazing Grace for Every Race" and opened the closed doors of a segregated church.

"We heard you," Henry said.

I hope they see, my children, that Christianity is much larger and more encompassing than the churches of my childhood or the youth groups of theirs, for all their goodness, much larger than any one sampling of the church universal, of that variegated tradition.

I think back to their baptisms at the cathedral in Baltimore. On my best days I allow the sacrament to relieve me of paternal anxiety for my children's soul care. Churchgoing and prayer before meals may not be their story; such practices may not be mine either, of course, these days. I think of baptism and I think of this faith as a river, and my children and I are all borne up by it, I believe.

———

I want you to know I finally read *Nausea*. I approached it the way I once would a country dog on an afternoon run, with all my assets, working every angle. I read the page silently and aloud; I purchased the book on audio and Kindle. I purchased a German translation and a UK edition with a daisy-dappled cover as sweet as a Laurie Colwin novel. During one difficult passage, I pressed the book down flat to the desk and underlined every sentence with the force of a branding iron. Eventually the back cracked and the pages scattered on the table like a flower's little seeds blowing hither. Other times I wrote out long passages by hand the way Japanese scholars once transcribed Faulkner novels in the special collections room at UVA.

It's an awful book; I understand why Camus damned it with faint praise. The ideas overflow the story. Its episodic unfoldings are tedious pleasures—and do not add up to a work of art. Sartre lingers over the repugnant features of humankind like a brooding Jansenist, like Pascal, he might have said, twisting and turning in the abyss. And yet its brooding feels staged. Where is there strength, sunlight, and desire? Where is Dan Aykroyd when you need him, existentialist half of the crime-solving team Sartresky and Hutch, who approaches the desperate war veteran with a pack of explosives strapped to his chest and asks, "Why does any man do what he does? Human reality is an action that means the determination is itself an action which reduces the action to a state of doing." Cut to smoke and debris of an exploding bomb.

———

Bonhoeffer wagered his life on the truth of the incarnate, crucified, and resurrected God, and thus on the conviction that following Christ leads to deep immersion in the world. "What remains for us is only the very narrow path, sometimes barely discernible, of taking each day as if it were the last and yet living it faithfully and responsibly as if there were yet to be a great future," he wrote shortly before he went to prison. I shall follow Bonhoeffer's example. I mean, will I?

———

I begin the day wiping blood after my morning stool and worrying about cancer. A doctor tells me it's nothing serious, but it's a sullen country nurse who gives me the brochure and delivers the verdict—anal papillitis. Let me tell you, my friend, that these words, once heard, are not easily forgotten. Inflamed anal crypts, the nurse said, and messy feces dribbling from a leaky anus, and how I should be relieved by the sight of bright red blood. Seriously?

In the safety of my car, I open the pamphlet. The page on which I land reveals only the benefits of a two-finger rectal dab of Vaseline, before and after a bowel movement, with illustrations by a sketch artist clearly influenced by the hyperrealist school. If I continue to find leakage in my underwear or in the seat of my pants, I might give an adult diaper a go. "It's time to break the stigma around incontinence products." Only when the blood runs dark should I worry.

Later that day, with the prescription filled and my office door latched shut, I struggle with the suppository. A website suggests that I squat over a small mirror that I've placed on the floor, positioned to facilitate a clean insert. Another advises a surgical glove generously lubricated. But the suppository splinters on impact, as does the tip of the glove—leaving me to reckon with a bare index finger wormed into my frightened ass. Oh, sweet Jesus and the angels.

———

Three days later, I open the Bible to the Gospel of Luke. Buxtehude's *Trio Sonatas* spins in the CD player, their exquisite textures and hues. I have been listening to him often of late and thinking again of Bonhoeffer's meditations on polyphony. "God wants us to love him eternally with our whole hearts," he wrote to Eberhard, "not in such a way as to injure or weaken our earthly love, but to provide a kind of *cantus firmus* to which the other melodies of life provide the counterpoint. . . . Where the *cantus firmus* is clear and plain, the counterpoint can be developed to its limits." Buxtehude's lines are straight, and you can hear what Bach heard and turned into the great canon of Western music, but inside those straight lines are hints of melancholy vernacular that are enlivened at times by notes of joyful defiance—sometimes I think it could have come out of Appalachia.

But at daybreak on the first day of the week, the women took the spices they had prepared and went to the tomb.

231

They found the stone rolled away from the tomb; but when they entered, they did not find the body of the Lord Jesus. While they were puzzling over this, behold, two men in dazzling garments appeared to them. They were terrified and bowed their faces to the ground. They said to them, "Why do you seek the living one among the dead? He is not here, but he has been raised. Remember what he said to you while he was still in Galilee, that the Son of Man must be handed over to sinners and be crucified, and rise on the third day." (Luke 24:1–7)

I want to stand with the women, inclined in astonishment. Though I'm not sure what's happening. Whether it's time interrupted or redeemed, or are the two the same? The stone is rolled away, the tomb is empty. Hoping in the resurrection of the dead as attunement elevated to creature-love and precision. Not so much living in mystery as living astonished, in view of being—astonishment at the groaning of creation, for a world made whole. Astonishment at hope for *others*. That the murdered children for whom we prayed around the children's altar in Baltimore might be awakened into love. That my Nana might see Mrs. Hamer in her radiant glory. That I might finally be freed from the vanity of my suspicions, so I might learn to love myself as God loves.

Still, fears of the afterlife remain fossilized in my imaginings. How can anyone feel certain that God wants us to taste the sweetness of the vine?

———

It's pleasant here in the late hours. Lights burn in the student apartments; such wakefulness is reassuring. I have the sudden notion that I want to think more about liturgy, prayer, and the Eucharist as a window raised to the sounds of a summer night.

There's a word for anxiety about fatigue. Some days begin here: how many hours or minutes before the mind fogs and eyes grow heavy, beckoning a nap. It's easy to imagine the dying body as a slow sink into nothing. Most days it hurts to work. I need surgery on my left hand. An old sports injury periodically shoots lightning bolts down both legs. I know the dying animal. Every few nights I carry Ginger up the stairs to her bed, because of the onset of degenerative myelopathy. Only rarely do I look down from the walking track at the AFC and wish I could still run full court with the kids. I'd rather lie down on my daybed, plug in my Bose earbuds, and listen to Gold Connections.

It's hard for me to recall the man who once took a lusty jog before lunch. Thirty minutes, sometimes longer. I would run around the circle of the UVA track. On warm days, I might bring a stopwatch and time myself on the 440. Foolish is the aging jock who wakes up of a morning in middle age and thinks he can break his personal record.

Could you give me once again the sight of my sons swimming in Ely's Harbour, as I sat in the shade and read *The Seven Storey Mountain* and *Waiting for Snow in Havana*?

———

To grow calm in the evening is a gift. To forget what was lost to fear.

A gift in its purest form presumes the unconditional. Gratitude is its only requirement. But gratitude has consequences. *Amen*s dust into footprints. This is the thing I've been trying to figure out for years. Immersed in the beautiful, broken earth, to reach for tenderness as if it were a cape jasmine shooting through an iron fence. Loved, in other words.

Acknowledgments

To and for
Karl Ackerman and Van Gardner (of blessed memory); Chris Coble, Kim Curtis, David Dark, Jessicah Duckworth, Carlos Eire, Ralph Eubanks, Christy Fletcher, Jon Foreman, Sarah Fuentes, Mark Gornik, Patricia Hampl, Susan Holman, Megan Hustad, Alan Jacobs, Chuck Mathewes, Mickey Maudlin, Isaac May, Alex Morris, Kristopher Norris, Dan O'Neill, Jessica Seibert, Darcey Steinke, Gregory Thornbury, Chantal Tom, Shea Tuttle, Nathan Walton, Wim and Donata Wenders, Brook Wilensky-Lanford, Lauren F. Winner; Mom, Dad, and mental health professionals everywhere; Karen, Henry, Will, and Nan ~

Thank you.

Notes

Martin Luther on Prozac

4 "I am like ripe shit": Martin Luther, *Tischreden* (Weimarer Ausgaben) V, no. 5537. Quoted in Erik H. Erikson, *Young Man Luther: A Study in Psychoanalysis and History* (New York: Norton, 1958), 206.

Harvard Divinity School: Fall 1981

11 I've often recalled the lesson: This paragraph is adapted from my memoir *The Last Days* (New York: Basic Books, 2001), 55.

12 "a muscular Norwegian": Elisabeth Elliot, *Secure in the Everlasting Arms: Trusting the God Who Never Leaves Your Side* (Grand Rapids: Revell, 2020), 14.

14 "I was moving in a": Diane Ackerman, *One Hundred Names for Love* (New York: Norton, 2011), 152, 153.

19 "Affliction is regarded": Dorothee Soelle, *Suffering*, trans. Everett R. Kalin (Philadelphia: Fortress Press, 1975), 17–18.

23 "across spaces so vast": Albert Camus, "The Sea Close By," *Personal Writings*, trans. Ellen Conroy Kennedy and Justin O'Brien (New York: Vintage, 2020), 191.

25 "The 'therapeutic movement'": Os Guinness, "America's Last

Men and Their Magnificent Talking Cure," in Os Guinness and John Seel, eds., *No God but God: Breaking with the Idols of Our Age* (Chicago: Moody, 1992), 122.

Dry Leaves Tumble Down University Circle

33 A research team at Baylor University: Christopher G. Ellison et al., "Prayer, Attachment to God, and Symptoms of Anxiety-Related Disorders Among U.S. Adults," *Sociology of Religion* 75/2 (2014): 208–33.

On Fire

47 "An asthma attack feels": John Updike, *Self-Consciousness: Memoirs* (New York: Random House, 1989), 94.

49 "the intelligence of the colored people": David M. Hargrove, *Mississippi's Federal Courts: A History* (Jackson: Univ. Press of Mississippi, 2019), 220–21; Frank R. Parker, *Black Votes Count: Political Empowerment in Mississippi After 1965* (Chapel Hill: Univ. of North Carolina Press, 1990); Charles V. Hamilton, "Southern Judges and Negro Voting Rights: The Judicial Approach to the Solution of Controversial Social Problems," *Wisconsin Law Review* 65 (Winter 1965): 72–102 (87).

53 We fashioned our own lingua franca: See John Dittmer, *Local People: The Struggle for Civil Rights in Mississippi* (Urbana: Univ. of Illinois Press, 1994), 58.

57 So I equipped myself for battle: The discussion of the unpardonable sin, my mother's concerns about ruined girls, and the canon of Christian sex ed my parents provided me with is lightly adapted from my book *The Last Days* (New York: Basic Books, 2001), 67–69; for a more extensive discussion of this season, see the chapter "The Joy of Fundamentalist Sex."

The Pursuit of a Literary Life

59 "Like Ireland's, Mississippi's history": W. Ralph Eubanks, *A Place Like Mississippi: A Journey Through a Real and Imagined Literary Landscape* (Portland, OR: Timber Press, 2021), 14–15.

59 "When you have built": Cited in Jerry Mitchell, "State's Contrasts Reach to Its Soul," *Jackson Clarion-Ledger*, February 14, 1994; also cited in Charles Marsh, *God's Long Summer* (Princeton, NJ: Princeton Univ. Press, 1997), 205.

71 "turning shadow into transient beauty": T. S. Eliot, "Burnt Norton," *Four Quartets* (New York: Harcourt, 1948).

HDS, Redux

77 "The reducto absurdum": William Faulkner, *The Sound and the Fury* (New York: Penguin, 1964), 95.

81 "era of good feeling": The phrase "era of good feeling" is Gil Troy's, in *Morning in America: How Ronald Reagan Invented the 1980s* (Princeton, NJ: Princeton Univ. Press, 2005).

83 "someone to unscrew my head": Cathy Park Hong, *Minor Feelings* (New York: Penguin, 2020), 4.

84 "When one is young": Friedrich Nietzsche, *Beyond Good and Evil: Prelude to a Philosophy of the Future*, trans. Walter Kaufmann (New York: Random House, 1966), 43.

85 *Krankheitsphase*: Erik H. Erikson, *Young Man Luther: A Study in Psychoanalysis and History* (New York: Norton, 1958), 27.

85 "the years in which we are most interested": Erikson, *Young Man Luther*, 27.

85 "every sensation is oppressive": George Eliot, *George Eliot's Life, as Related in Her Letters and Journals*, ed. J. W. Cross (New York: Houghton Mifflin, 1908), 269.

Christian Anxiety: A Short Theology

89 "a blood-red sunset": Kay Redfield Jamison, *An Unquiet Mind: A Memoir of Moods and Madness* (New York: Random House, 1995), 80.

89 "The smoke is an animal": Hans Urs von Balthasar, *The Christian and Anxiety,* trans. Dennis D. Martin and Michael J. Miller (San Francisco: Ignatius, 2000), 70–71.

90 "Anxiety ennobles": Søren Kierkegaard, *The Concept of Anxiety*, trans. Reidar Thomte (Princeton, NJ: Princeton Univ. Press, 1980), 155.

90 "the deity's creative birth pangs": Søren Kierkegaard, *The Concept of Anxiety*, trans. Alastair Hannay (New York: Liveright, 2014), 72.

89 "The crucified God is near": Jürgen Moltmann, *Der gekreuzigte Gott: Das Kreuz Christi als Grund und Kritik christlicher Theologie* (Munich: Christian Kaiser, 1972), translation mine.

Charlottesville: The First Sojourn

96 "the irksome particulars": I borrow this excellent phrase from the poet Matt Cook's *Irksome Particulars* (Atlanta: Publishing Genius, 2019).

103 "it is only when one loves": Dietrich Bonhoeffer, *Letters and Papers from Prison*, ed. Eberhard Bethge (New York: Simon & Schuster, 1997), 157.

Cathedral Light

121 "Religion has the potential": Peter Pressman, John S. Lyons, David B. Larson, and John Gartner, "Religion, Anxiety, and Fear of Death," in *Religion and Mental Health*, ed. John F. Schumaker (New York: Oxford Univ. Press, 1992), 98–109.

125 "they believe that conditions": Kathryn Joyce, "The Rise of Biblical Counseling," *Pacific Standard*, September 2, 2014, https://psmag.com/social-justice/evangelical-prayer-bible-religion-born-again-christianity-rise-biblical-counseling-89464.

126 "competent to counsel": Joyce, "Rise of Biblical Counseling."

126 "Biblically, there is no warrant": Jay E. Adams, *The Christian Counselor's Manual: The Practice of Nouthetic Counseling* (Grand Rapids, MI: Zondervan, 2010), 9.

127 "O be quiet!": Joyce, "Rise of Biblical Counseling."

130 Birds pecking . . . aborted fetuses: See Alan A. Bernstein, *The Formation of Hell: Death and Retribution in the Ancient and Early Christian Worlds* (Ithaca, NY: Cornell Univ. Press, 1997), 288, passim.

132 "spellbound . . . through a forest of memories": Hermann Hesse, *Beneath the Wheel* (London: Picador, 2003), 31.

141 "The question remains open": Paul Ricoeur, *Freud and Philosophy: An Essay on Interpretation* (New Haven, CT: Yale Univ. Press, 1977), 235.

142 "a child of the nineteenth century": Paul Tillich, *The Courage to Be* (New Haven, CT: Yale Univ. Press, 1952), 81. Tillich is the rare modern Protestant theologian who grappled with anxiety and extolled the benefits of psychoanalysis. But his accounts linger amid ontological abstractions; you will not find any consideration of anxiety's harsh somatic presences or clinical context—or his own tortured sexuality.

142 "funerary sermon": Robert Metcalf, "The Word of Freud: Our God Is Logos," *Journal for Culture and Religious Theory* 3/2 (Spring 2002), https://jcrt.org/archives/03.2/metcalf.shtml.

142 "the power of creating": Tillich, *The Courage to Be*, 81.

Outtakes from an Evangelical Analysis

149 "Say men what they will": John Robinson, *New Essays: Or, Observations Divine and Moral* (British Library, 1625), 308–9.

149 "The Bible speaks": Sarah Boesveld, "Q & A with Michael Pearl," *National Post*, November 12, 2011, https://nationalpost.com/news/qa-with-michael-pearl.

150 When asked if controversy: Boesveld, "Q & A with Michael Pearl."

150 "Corporal chastisement is not": Anderson Cooper interview with Michael Pearl, *Anderson Cooper 360 Degrees*, October 26, 2011, https://transcripts.cnn.com/show/acd/date/2011-10-26/segment/02.

151 American practices of corporal punishment: See Donald Capps, *The Child's Song: The Religious Abuse of Children* (Louisville: Westminster John Knox, 1995).

155 Desire is sorrow in my heart: See Walker Percy, *The Moviegoer* (New York: Knopf, 1961), 68.

155 "comes howling down Elysian Fields": Percy, *Moviegoer*, 228.

159 "We can't imagine": Ezra Klein, "Rebecca Traister on the Coming #MeToo Backlash," *Vox*, December 20, 2017, https://www.vox.com/2017/12/11/16756158/rebecca-traister-sexual-harassment-trump-clinton.

159 last lines of *Middlemarch*: Viz., "the growing good of the world is partly dependent on unhistoric acts; and that things are not so ill with you and me as they might have been is half owing to the number who lived faithfully a hidden life, and rest in unvisited tombs."

160 "work with girls who will": Lillian Smith, quoted in Ann Short Chirhart and Kathleen Ann Clark, eds., *Georgia Women: Their Lives and Times*, vol. 2 (Athens: Univ. of Georgia Press, 2014), 175.

Summer in Laurel

163 "Where analyses in Freud's early days": Daniel Pick, *Psychoanalysis: A Very Short Introduction* (Oxford: Oxford Univ. Press, 2015), 75.

167 "effloresce of new insights": Peter Gay, *The Bourgeois Experience: Victoria to Freud, vol 1, The Education of the Senses* (New York: Norton, 1984), 280.

170 God "can stoke [us] with madness": Robert Burton, *The Anatomy of Melancholy*, ed. A. R. Shilleto (London: G. Bell and Sons, Ltd., 1920), 203–204.

173 "my presence runs counter to their interests": Joan Didion, *Slouching Toward Bethlehem* (New York: Farrar, Straus, and Giroux, 1968), xvi.

Years of Wonder and Longing

184 "being reliably present": D. W. Winnicott, *The Family and Individual Development* (New York: Routledge, 2006), 44.

184 "The father may be absent": D. W. Winnicott, *Home Is Where We Start From: Essays by a Psychoanalyst*, ed. Clare Winnicott, Ray Shepherd, and Madeleine Davis (New York: Norton, 1990), 132.

Depression

190 "Faith means seeking not noise": Karl Barth, *The Word of God and the Word of Man*, ed. Douglas Horton (Gloucester, MA: Peter Smith, 1978), 26.

192 she had prescribed medicine routinely for patients: Julia Kristeva writes: "Psychoanalysis wagers to modify the prison of the soul that the West has made into a means of survival and protection, although this prison has recently been revealing our failings. This wager is therapeutic as well as ethical, and incidentally, political. Yet although we may seek the acceptance and even expansion of psychoanalysis, our wish is coming up against some substantial barriers." Then she goes on to talk about needing to pay attention to new research in psychopharmacology ("In Times Like These, Who Needs Psychoanalysis?" in *New Maladies of the Soul*, trans. Ross Guberman [New York: Columbia Univ. Press, 1995], 29).

193 "can reestablish conductivity among neurons": Ross Mitchell Guberman, ed., *Julia Kristeva Interviews* (New York: Columbia Univ. Press, 1996), 87.

194 "wastes and deserts of the soul": Virginia Woolf, *On Being Ill* (Ashfield, MA: Paris Press, 2002), 11.

The Grace of the Strong Sin

198 "through open gates": Martin Luther, Preface, *The Complete Edition of Luther's Latin Works*, vol. 1 (1545). Quoted in Alister E. McGrath, *Christian Theology: An Introduction*, 6th ed. (Malden, MA: Wiley Blackwell, 2017), 388.

198 "That, if we have no such thing": Martin Luther, *On the Bondage of the Will*, trans. Henry Cole (London: Bensley, 1823), 286.

199 "Christians talk about": Percy, *The Moviegoer*, 200–201.

200 "men tell lies vigorously": Dietrich Bonhoeffer, *Letters and Papers from Prison*, ed. Eberhard Bethge (New York: Simon & Schuster, 1997), 157.

200 "on the verge of some kind": Dietrich Bonhoeffer, *Dietrich Bonhoeffer Works,* vol. 16, *Conspiracy and Imprisonment 1940–1945*, ed. Mark Brocker, trans. Lisa E. Dahill (Minneapolis: Fortress, 2006), 329.

201 "concrete bodily experience": Dietrich Bonhoeffer, *Dietrich Bonhoeffer Works*, vol. 8, *Letters and Papers from Prison*, ed. John W. de Gruchy, trans. Isabell Best, Lisa E. Dahill, Reinhard Krauss, and Nancy Lukens (Minneapolis: Fortress, 2010), 448–49, 454, 456. See also Charles Marsh, *Strange Glory: A Life of Dietrich Bonhoeffer* (New York: Knopf, 2014), 372.

On Christian Counseling

205 "confront[s] every man nouthetically": Jay E. Adams, *Competent to Counsel* (Grand Rapids, MI: Zondervan, 1970), 42.

207 "utterly corporeal body": Teresa Brennan, *The Transmission of Affect* (Ithaca, NY: Cornell Univ. Press, 2004), 3.

201 "That's as it should be": For a fuller discussion of the story of Finkenwalde, see Marsh, *Strange Glory*, 227–62.

Oh, Merton

211 "Much more important are": Thomas Merton, *The Sign of Jonas* (New York: Harcourt, 1981), 8.

213 "The Christian Church is the greatest": Friedrich Wilhelm Nietzsche, *The Antichrist*, trans. Anthony M. Ludovici (Amherst, MA: Prometheus, 2000), 106.

213 "word bombardments": Kathleen Norris, *Dakota: A Spiritual Geography* (New York: Mariner, 2001), 94.

214 "homilophobia": psychtimes.com/homilophobia-fear-of-sermons/.

219 "I wish nobody had ever told me": Thomas Merton, *The Journals of Thomas Merton*, vol. 1, *Run to the Mountain: The Story of a Vocation* (1939–1941), ed. Patrick Hart, OCSO (San Francisco: HarperSanFrancisco, 1995), 96.

219 "The more you try to avoid suffering": Thomas Merton, *The Seven Storey Mountain* (New York: New American Library, 1948), 86.

Quiet Days in Charlottesville

220 "the anxiety that comes upon Jesus in waves": Hans Urs von Balthasar, *The Christian and Anxiety*, trans. Dennis D. Martin and Michael J. Miller (San Francisco: Ignatius, 2000), 71–76.

221 "Perhaps it might be said": Augustine, *Confessions*, 2nd ed, ed. Michael P. Foley, trans. F. J. Sheed (Indianapolis: Hackett, 2006), 245. See also page 243: "O, my Lord, my Light, here too man is surely mocked by Your truth! If we say the past was long, was it long when it was already past or while it was still present? It could be long only while it was in existence to be long. But the past no longer exists; it cannot *be* long, because it is not at all."

225 "lonely center of the spinning earth": Jane Stembridge, *I Play Flute and Other Poems* (New York: Seabury, 1966), 30.

225 "When we loved": Stembridge, *I Play Flute and Other Poems*, 128.

230 "What remains for us": Dietrich Bonhoeffer, *Dietrich Bonhoeffer Works*, vol. 8, *Letters and Papers from Prison*, ed. John W. de Gruchy, trans. Isabell Best, Lisa E. Dahill, Reinhard Krauss, and Nancy Lukens (Minneapolis: Fortress, 2010), 50.

231 "God wants us to love him": Dietrich Bonhoeffer, *Letters and Papers from Prison*, ed. Eberhard Bethge (New York: Simon & Schuster, 1997), 303.